蒋国兴 主编　　徐宾宾 策划

江苏凤凰科学技术出版社

以"十年"一叙为序

2012年在长白山,一次"大木设计"的活动上我与国兴相识。当时,一袭黑衣的他话虽不多却略透侠气。我们是在以江浙老乡对阳澄湖大闸蟹的共识中开始融洽了彼此的交流,餐饮空间设计专辑《叙品十年 蒋国兴作品集》即将面世,国兴力邀我为其作序。谈美食环境我不甚有信心,涉及就餐的情怀体验我认为还能说上几句。在设计界,像他这样有如此多元餐饮空间品牌的设计师确实不多,能为其作序甚是有幸,我们会感受到这位同仁对设计情怀的独特诠释。

南方雨后的湿润被一点点细腻地渗透到西域干燥的边寨,丝绸之路上的戈壁风沙也变甜了。

叙品十年,从昆山出发到融入新疆,越走越远,与之对应的设计理念则更趋纯净、更加坚守东方的回归,这与蒋国兴一贯的信念和追求密不可分。在叙品十年的作品中你会发现一个职业人让理想变为现实的清晰的目标需求。

几年来,我一直关注他的作品并不断发现,一路走来,他设计调性的变化及理念的纯粹让曾经的奢华已然褪去,留下更多的是触击本质,如"原膳""水云间""不净"和"叙品茶事"等作品。其中,"水云间·茶会所"会让人感受到诗与生活可以通过竹影在空间内摇曳完成体验,而"不净"却将黑砖和木纹应用到极致并用禅意诉说空间美学。在最新作品"叙品茶事"中,他又以最朴素的原料——土砖和红瓦阐述了一个自然的故事。真正进入自然并非要脱掉鞋子走在冰凉的田埂和山野,它的本质是回归到灵魂的初心。

当一个人精神、阅历丰富到一定程度,他是不需要显摆昂贵的材料或是抢眼的符号去掠夺你对他的惊讶的。真正做到对朴素追求的素养取决于他个人成长的经历和形成的独特价值判断。使用最平价的原材料,需要有相当的能力和技巧才能诠释不朴素的视觉联想……

在他许多的商业空间内,他能充分调度当今多媒体的演绎手法,将沉睡在书画作品中的意境立体化于空间中。当人们进入到这样的商业空间时,可以体验到与山水与诗意近距离的交融。国兴熟知江南文化的精致极简的表达,在设计中和谐地呈现在大山大河的辽阔下。回顾国兴这十年的作品,我欣赏这其中所蕴含的人文气息,我相信,有文化,才能经久流传,而浅薄,终究将被淘汰。

历来人们对光的感受只是停留在文字的表达上,而叙品空间中对光的运用和展示是具备预谋的。国兴善于使用白桦、枯竹和土砖等自然材料,一草一木,一树一竹,都能感受到他对传统文化及业主经营上的双重平衡,对环保的坚持和对品质的追求也是持之以恒的。黑、白、灰三种极度难用的颜色,却能被他在黑灰围抱中演绎得出神入化。

叙品十年的作品中,可以折射出他的创作思想对整个团队的覆盖和有效的执行渗透力,今年亦是国兴所创的"叙品设计"成立十周年,此时此刻,出版本书有着特殊的意义:既是对过去十年的回味,也是对下一个十年的期望。谨以此序对国兴及叙品设计表示祝贺。

陈耀光,2016年8月,杭州

陈耀光

室内设计师/生活艺术家/千岛湖岛主

1987年毕业于中国美术学院首届环艺系

十大最具影响力华人设计师

1996—2005年十届中国室内设计大赛七次一等奖得主

2010年获"亚洲五十摩登绅士"

2014—2016年连续受邀为米兰设计周参展设计师

Ad100中国榜单得主

创基金理事

杭州典尚建筑装饰设计有限公司创始人及创意总监

Preface on the "Decade"

In 2012, I met Guoxing at Damu design activity in Changbai Mountain. He was in black, and seems to be silent and justice. We started to talk for the common topic of Yangcheng Lake hairy crab which is a specialty in Jiangsu. The book of restaurant space design named "Xupin Decade—Jiang Guoxing Portfolio" is about to publishing. Guoxing invited me sincerely to preface it. I can't talk about dietary environment good enough, but I have some personal feeling for restaurant. In design circle, there are many designers like Guoxing who have various space bands of restaurant. I am honored to preface for his book, and we will see his unique explanation to emotional design.

After the rain in south, air moistens gradually dry region in west. The Gobi Desert becomes sweet in the Silk Road.

Xupin starts from Kunshan and goes further and further, and melts with Xinjiang culture well in recent decade. In the same way, its design idea gets closer to purity, to orient culture which is Guoxing's belief and pursuit all the while.

For several years, I kept in pay attention to his works, and I found that time makes him be close to nature and atman, instead of extravagance and vanity. Many works like "Raw Food", "Shuiyunjian•Tea Club", "Buzheng" and "Xupin Tea-things" are brilliant. The work "Shuiyunjian•Tea club" expresses the art about poem and life by hazy bamboo shadow. The work "Buzheng" expresses the art of Zen by the excellent application of black brick and wood grain. "Xupin Tea-things" is his latest work, he also chooses plain materials like brick and red tile and shows a natural story. The real way to near nature is not to take off shoes and walk on cool ridge or mountain road, but to find inner peace.

The one who has a rich spiritual world, doesn't need to impress people by expensive material or shiny symbols. The one who lives plainly always has experienced a lot and has unique, characteristic sense of world. It is not easy to show splendid visual space through simple and plain materials.

In many commercial space that he designed, he applied multimedia technology with landscape elements in it. People experience the art of poem and nature when they are there. Guoxing is good at refining the Jiangnan culture, and applying it to space design. His works in recent decade send forth rich cultural fragrance. I am in favor of retaining of culture and elimination of vanity.

People feel the light by the way of literature all along. Xupin has plans for light applying. Guoxing is skilled in the use of white birch, dry bamboo and brick etc, and keeps a balance between traditional culture and commerce. It is worth to persist to protect environment and pursue quality. The color of black, white and grey are difficult to match in harmony, but he does it well.

We can feel that his idea is executed by his team effectively in the works over the past decade. This year is also the 10th anniversary of "Xupin Design" set up by Guoxing. At the moment, publishing the book means to take over from the past and set a new course for the future. Congratulations!

Chen Yaoguang

August 2016, Hangzhou

编者心语

酝酿多年的《叙品十年 蒋国兴作品集》（以下简称《十年》）终于面世，内心既忐忑又兴奋。在电子书盛行的今天，叙品选择以纸质书出版，不是逆势以彰显个性，而是想以此书来记录我们曾走过的十年，也作为十周年庆典纪念品分享给喜欢叙品原创设计的朋友们。

从2006年的三个人到现在近百人的设计团队，叙品经历了很多设计公司未曾经历过的酸甜苦辣、风风雨雨。有各种辛酸，也有成功的喜悦和快乐的泪水。但无论经历过什么，我们初心不变："我们只做原创设计"早已根植于每位叙品人的心。

"时间让一切变得更有意义"，而十年沉浮让叙品更懂得珍惜和感恩：珍惜得到的一切，感谢帮助我们成长的朋友。失去的如果努力还会再次得到，而得到的如果不珍惜也会再次失去，认真做事，开心做人。如今的我们不紧不慢，学会以豁达的心态包容一切，以积极的态度去应对未知的未来，我们相信，在下个十年一切也会因此而变得更有意义。

《十年》收录了叙品过去十年较为典型的25个案例，其中隐约可以发现叙品多年来一直研究探索的现代中式"无物"主义的影子。也许这种思想还只是在初步阶段，但庆幸的是我们已经在路上，也希望自己能在这条路上越走越远。

很荣幸能邀请到在设计界享有无数荣誉并广受业界尊重的陈耀光老师为《十年》作序。陈老师能在百忙中为本书点评，是我的荣耀，也是叙品的荣幸。陈老师对叙品一直以来的鼓励与支持是叙品坚持原创的动力，在这里我们表示无比的感谢。

谨以此书献给我深爱的家人以及所有喜欢原创和尊重原创的朋友们！

蒋国兴
2016年9月

Author's Note

It is nervous and exciting that the book "Xupin Decade—Jiang Guoxing Portfolio" (In short "Decade") prepared for many years is about to be published. E-book is popular in the present, but the purpose of that Xupin chooses to publish works by the traditional way of paper printing is to record the experiences for ten years, and consider the book as a souvenir for 10th anniversary sharing to friends who like Xupin, instead of showing personality by swimming against the tide.

From three persons in 2006 to now nearly a hundred people in the team, Xupin has been through all the ups and downs, all the laughter and tears that many design firms have not gone through. But whatever we have experienced, everyone in Xupin insists on the principle of making original design.

"Time makes everything significant", all the ups and downs for a decade make Xupin understand the importance of cherishing and thankfulness. To cherish everything we got, to thank everyone who helps us.

There are 25 typical cases for over ten years gathered in "Decade". We can find that Xupin has been keeping exploring the presentation about "nothing" in modern-chinese style. It is possibly tentative, but it's fortunate that we have already been on the road and we shall move on.

It's an honor for me and Xupin for inviting Professor Chen GuangYao who is respected and gets countless reputation in design circle to preface for "Decade". We really appreciate his encouragement which is driving source for insisting on original design.

This book is dedicated to my beloved family and friends who love and respect original works.

Jiang Guoxing

September 2016

CONTENTS / 目录

008
Shuiyunjian · Tea Club
水云间 · 茶会所

022
Lan Xiu Restaurant
揽秀食府

034
Puyu Exhibition Room
璞玉展厅

042
Delicious Soup House,
Thailand Seafood Chafing Dish
靓汤房子 · 泰式海鲜火锅

050
Meiya IMAX Cinema
美亚巨幕电影院

064
No.1 Deep Sea
深海壹号

076
Drunk Yue Restaurant
醉玥餐厅

086
New Air Health Management Center
新空气健康管理中心

096
Longhai Construction Engineering Group, Suzhou
苏州龙海建工

110
Xupin Design Suzhou Corporation
叙品设计苏州公司

122
Buzheng Vegetarian Restaurant
不诤素食馆

136
Hua Shi Jian Restaurant
花食间

150
The World Peacock Restaurant
般若世界 · 孔雀餐厅

160
No.1 Zhuxi, Changchun Road
竹溪一号长春路店

172
Sanqian Health
三仟健康

182
Sha Wei Legend
沙味传奇

196
The Reception Center of Millennium Beauty Cosmetic Surgery Hospital
千禧丽人整形医院接待中心

206
Unity Tea
合一茶道

218
Horizon Club
天域阁

232
Flowers-boiled Fish
花枝沸腾鱼

242
Chinatown Elegant Foot Porch
唐人街雅足轩

258
June Henderson Office
君元恒基

270
Raw Food
原膳

282
Yunfeng Investment Company
云峰投资公司

296
Songshan Xing Foot Massage
松山行足道

水云间·茶会所
Shuiyunjian•Tea Club

项目名称：水云间·茶会所
项目地点：新疆乌鲁木齐
建筑面积：460 ㎡
主要材料：黑色花岗岩、火山岩、海藻泥、鹅卵石、米黄洞石
主案设计：蒋国兴
空间摄影：吴辉（牧马山庄空间摄影机构）

Project name: Shuiyunjian · Tea Club
Project location: Urumqi, Xinjiang
Building area: 460 ㎡
Main materials: The black granite, volcanic, algae clay, cobble, beige-cream travertine
Project designer: Jiang Guoxing
Photograph: Wu Hui (Graze horse villa photograph organization)

因为参加了视觉空间"本生行"改变了自己多年来的工作态度，同时也收获了一个自己喜欢的案子——水云间·茶会所，在此由衷感谢"本生行"。

水云间的设计过程异常顺利，从北京飞乌鲁木齐的飞机上，用2个小时就完成了方案，但接下来与设计师的沟通用了5个小时，与业主的沟通用了2个小时，表现图前后修改了三次，也是本人第一次让设计师修改表现图，颇感抱歉。

以下三段是水云间设计的关键语：

1. 周敦颐读书修行的月岩洞，大自然鬼斧神工地留下如此震撼之天坑，拾级而上，头顶云彩飘飘，平步青云的那种喜悦感油然而生。

2. 曾国藩故居也许是本人见过的最喜欢的中式建筑，明式的纤细做了加法，清式的细节做了减法。导游说过：过低的八角门来形容为官的低人一等及生意上的八面玲珑！我深深地将此记在心中。餐厅桌椅底下一面镜子反映了此物必是曾家的细节，也稍稍让人感动。

3. 徐志摩的美文：轻轻的我走了，正如我轻轻的来；我轻轻的招手，作别西天的云彩。那河畔的金柳，是夕阳中的新娘；波光里的艳影，在我的心头荡漾。软泥上的青荇，油油的在水底招摇；在康河的柔波里，我甘心做一条水草！那榆荫下的一潭，不是清泉，是天上虹；揉碎在浮藻间，沉淀着彩虹似的梦。寻梦？撑一支长篙，向青草更青处漫溯；满载一船星辉，在星辉斑斓里放歌。但我不能放歌，悄悄是别离的笙箫；夏虫也为我沉默，沉默是今晚的康桥！悄悄的我走了，正如我悄悄的来；我挥一挥衣袖，不带走一片云彩。

I sincerely appreciate "Ben Sheng Hang", for the reason that I have changed my attitude toward my job by attending Visual Space "Ben Sheng Hang", and gotten a good project- "Shuiyunjian·Tea Club" at the same time.

The design process of the project is smooth, I completed the project for 2 hours on the flight from Beijing to Urumqi city, then I communicated with the designer for 5 hours, with the owner for 2 hours. During the whole process, the drawing was modified in three times. This is the first time I let the designer modify the drawing, I feel so sorry…

The following three sections are the keywords of the project:

1. The Yue Yan hole where Zhou Dunyi did academic research is a miraculous natural pit. People will see charming fluttered cloud and feel successful on social life.

2. The former residence of Zeng Guofan is probably my favorite Chinese architecture that I had ever seen. It simplifies the slender elements of Ming-style, and enriches the details of Qing-style. The local guide says: "The short eight-angle door suggests humility of a statesman and the flexible of a businessman. The detail that the mirror under the dinner desk symbolizes the possession of owner is impressive."

3. A poem of Xu Zhimo:

Very quietly I left

As quietly as I came here;

Quietly I wave good-bye

To the rosy clouds in the western sky.

The golden willows by the riverside

Are young brides in the setting sun;

Their reflections on the shimmering waves

Always linger in the depth of my heart.

The floating heart growing in the sludge

Sways leisurely under the water;

In the gentle waves of Cambridge

I would be a water plant!

That pool under the shade of elm trees

Holds not water but the rainbow from the sky;

Shattered to pieces among the duck weeds

Is the sediment of a rainbow-like dream?

To seek a dream? Just to pole a boat upstream

To where the green grass is more verdant;

Or to have the boat fully loaded with starlight

And sing aloud in the splendour of starlight.

But I cannot sing aloud

Quietness is my farewell music;

Even summer insects help silence for me

Silent is Cambridge tonight!

Very quietly I take my leave

As quietly as I came here;

Gently I flick my sleeves

Not even a wisp of cloud will I bring away.

一茶一世界，一味一人生。人生如茶，第一道茶苦若生命、第二道茶香似爱情、第三道茶淡如清风。在这个飞速发展、喧闹的城市中，寻求内心一份真实的平静，茶会所便是首选之地。本案用现代中式风格展示了一个素雅、别致的茶文化空间。在平面布置上分为两层，一楼规划了大厅、接待区、包间、景观区，二楼规划了4个小包间和1个大包间、厨房、卫生间、办公室、库房等。在色彩运用上，以黑色、白色、做旧木本色为主色调，灰色为辅色调。

进入前厅首先看到的是水泥块整齐堆砌的背景墙，每一个洞口都摆满了蜡烛灯，散发出若隐若现的灯光。穿过锈铁板的移门，便是大厅。锈铁板上雕刻的不规则"水云间"字体悬挂于火山岩墙面上。墙面的柜子被分成了大大小小的框架，每一个框架又被等分成许多小格子，很像古代药房的小柜子。内凹的小格子内摆满了各种各样的小陶罐，在淡黄色灯光的映照下愈发显得精致。

前厅的后面是接待区，没有奢华的装饰，而是以一面残岩断壁土墙阻挡了与大厅间的交流。右边是景观区和包间。枯竹、石头、云灯装饰的景观，无论你置身在室外还是室内的入口处、接待区、大厅，还是包间，都能欣赏到，可谓点睛之笔。包间与大厅之间没有完全隔开，玻璃隔断很好地保留了与大厅间的视线交流。火山岩装饰的墙面挂着竹篱笆，木制花格装饰的柜门、黑白的挂画，无一不透露着中式情节。

A cup of tea is a world, a kind of flavor is life. Life is like tea. The first course of tea is the same bitter as life, the second course is like love, and the third course is the same light as breeze. In this rapidly developing and noisy city, if you want a real and peaceful mood, the tea club is the best choice.

A TV background wall in the front hallway, which is piled by cement blocks. Some candle lamps that emanate soft light are set in holes. Going through the moving door which is made of the rusted plate, you get to the lobby. Irregular letters "Shuiyunjian" which curved on rusted plate are hung on the volcanic wall. The cabinet on the wall is separated to frames of all sizes, and every frame is separated to grids of all sizes. This is like the medicine-chest in ancient China. In these inset grids, there are all kinds of small steans, which are very delicate in the canary yellow light.

The reception area behind the front hallway has no luxurious decoration. There is only a wall to block the communication with the hall. On the right side, it's the landscape area and private rooms. You can appreciate dry bamboo, art stone, cloud lamp decoration wherever you are at entrance or in reception area, lobby or private room. It's a wonderful point in the whole design. The private room is not separated completely with lobby; glass partition keeps the visual communication between the two spaces. Some bamboo fences are hung on the volcanic wall, some cabinet doors decorated with wood grillwork, and some black and white paintings, all of which reveal the sense of Chinese-style.

015

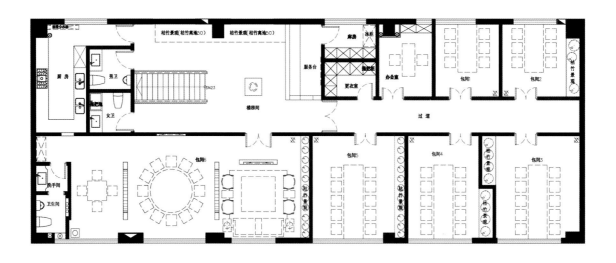

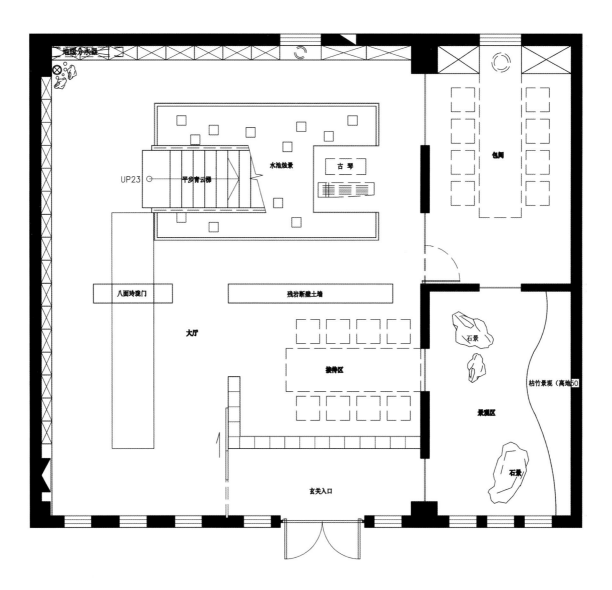

　　在通往楼梯的入口处，设计师运用中国古典元素，采用迂回渐进的设计手法，特意设计了一个八角隔断墙，意味着人生的八面玲珑。一条长长的桌子穿过八角立在那里，桌面的下面是一块不规则的锈铁板，背面写了几首诗词，通过黑色亮面的石材反射到地面上，给黑色的地面增加了一点点色彩。桌子的上面悬挂着一盏竹竿装饰的吊灯，既特别又实用。绕过隔断墙，一个巨大的水池置于中央，一边是楼梯的入口，直通二楼，顶面还飘来一片云朵，意味着生意上的平步青云。另一边是悬空于水池上方的古筝演奏区。水池中规划了很多蜡烛灯，好像许多蜡烛漂在水面上一样，柔弱的灯光映在水面上，再配上古筝的演奏声，既和谐又美好。通往二楼的楼梯简洁实用，黑色的踏步、白色的栏杆、古铜色的扶手、暗藏的灯带、顶面的云灯，墙面上则有徐志摩的诗词："轻轻的我走了，正如我轻轻的来"。

　　At the entrance to the stairs, the designer uses some Chinese classic elements to design an eight-angle grillwork with the circuitous and gradual design method, which means life is auspicious. A long desk through the grillwork stands there, under the desk there is an irregular rusted plate, at the back carved some poems which are reflected onto the ground to make the ground more colorful. A droplight decorated with bamboo is special and useful. Passing the grillwork, a giant pool is set in the center. On the one side, it is the entrance of stairs, going through the second floor, a cloud decoration is floating over the head, which means to have high status on social life. On the other side, it is the Gushing playing area hung over the pool. There are some candles lights designed in the pool, which looks like many candles float on the water. It must be a peaceful feeling when you listen to the music and stare the soft light. The stair going through the second floor is concise and useful, and there are the black footstep, the white railing, the bronze handrail, the hidden LED lamp, the cloud lamp on the top, the poem of Xu Zhimo on the wall: Very quietly I left, as quietly as I came here.

二楼的左边是竹子装饰的景观，木拼条装饰着服务台及整个走道。走道没有多余的光源，三条回字形的灯片把过道分成一段一段的。在每个暗门的入口处都悬挂了一个锈铁板雕刻的门牌，一束束光源照射在每个门牌上。在走道的尽头，一盏云灯点缀着整个过道，墙面的镜子拉伸了空间感。

包间内没有复杂的造型，以素色为主，米黄洞石、鹅卵石、灰色木拼条、灰色火山岩装饰的墙面，再搭配竹篱笆、黑白挂画、枯枝等装饰品。在每个包间都规划了一处竹子的景观，使空间更具有情调。

卫生间延续了走道的元素，木拼条装饰的墙面、黑色的地面使空间看起来更质朴。卫生间延续了走道的元素，木拼条装饰的墙面、黑色的地面使空间看起来更质朴。

On the left side of the second floor is bamboo decoration, and some joint battens decorate service counter and corridor. There is no more light in corridor, three lights in a shape of "回" (a Chinese character) cut the corridor into sections. At the entrance of every hidden door hang a doorplate curved by rusted plate which is lighted up by a beam of light resource. At the end of corridor, a cloud lamp embellishes the whole corridor, and the mirror on the wall extends the sense of space.

The private room decoration is concise with plain color, the wall of which decorates with all kind of decoration materials like beige travertine, cobbles, gray joint battens, and gray volcanic, and matches some bamboo fences, black and white paintings, dry branches. Every private room has a bamboo artwork, which makes the space more emotional.

The washroom continues the element of corridor, battens and black ground make space plain.

揽秀食府
Lan Xiu Restaurant

项目名称：揽秀食府
项目地点：新疆乌鲁木齐
建筑面积：1550 ㎡
主要材料：爵士白大理石、锈铁板、藤编壁纸、彩色玻璃、拼花地砖等
主案设计：蒋国兴
空间摄影：吴辉（牧马山庄空间摄影机构）

Project name: Lan Xiu Restaurant
Project location: Urumqi, Xinjiang
Building area: 1550 ㎡
Main materials: Jazz white marble, rusted plate, rattan wallpaper, colorized glass, pattern combination tile
Project designer: Jiang Guoxing
Photograph: Wu Hui (Graze horse villa photograph organization)

　　本案定位为大众化的百姓消费场所，同时体现一定的文化主题，为餐厅的经营赋予更有层次的文化内涵。因此在环境和空间设计上充分考虑其独特性和新颖性，体现品牌的特点和优势以吸引更多消费者的光临。

　　空间以重庆的传统文化为基调，以千年古镇重庆缩影的手绘画为形象元素。本案在设计中融合了雾都古镇的建筑特色，并将当地百姓朴实的生活或民俗物品陈设于空间之中，使就餐者无形中置身于老重庆的文化氛围之中，既享受了美食又领略了老重庆文化的魅力。整个空间分为两层，一楼主要是前台接待区、散座区、卫生间，二楼是高档包间区。

The project is located at civilian consumption place in a popular style, and reflects cultural theme in some degree. It makes the cooking culture of restaurant get a higher level. The design of the whole atmosphere and space is considered creatively to reflect characters of brand and attract more customers.

The keynote of space is Chongqing's traditional culture, applying elements of hand drawing of Chongqing – a millennium town. The project design blends in some architecture characters of this foggy town and displays the folk custom objects of the local life. It makes customers enjoy charm of food and feel cultural atmosphere of old Chongqing city. The whole space is divided into two floors. On the first floor it includes reception counter, relax zone, washroom, and on the second floor there are some luxury private rooms.

电梯厅没有复杂的造型，中国黑荔枝面的石材装饰着墙面，挂着重庆建筑特有的黑白挂画，旁边放上原木色的条凳，土黄色的小陶罐插着几枝花，地面采用拼花小地砖。

接待区的背景是蘑菇石装饰的墙面，锈铁板的层板陈列着民间生活器具——陶罐营造生活的意境。锈铁板的墙面悬挂着"雾之都"的标志，既突出了餐厅的名字，又给人一种耳目一新的视觉感受。几块实木板组合而成的服务台，简单又大气。

The decoration in the elevator hall is concise. The wall is decorated with black bush-hammered stone, and hangs the black and white paintings of Chongqing architectural characters, burly wood bench, little wheat steam with a few flowers. Mosaic tiles are paved on the ground.

The background of reception desk is a beautiful wall decorated with mushroom stones. Rusted plates displaying local life-style steans create a flavor of living. The logo of "雾之都"(Chinese characters) is hung on the rusted wall, which represents restaurant's name and makes people feel creative. The service desk combined with boards is concise and advanced.

包间的沙发背景是一幅巨大的千年古镇重庆缩影的黑白手绘画，墙面采用藤编壁纸加木作小线条，做旧的中式装饰柜，铁艺的吊灯，无一不体现着老重庆的文化特色。

公共卫生间延续了走道地面的元素，拼花小地砖铺满了整个卫生间的墙面和地面，红色的柜子给人眼前一亮的感觉。包间卫生间则采用了白色的色调，爵士白的大理石，搭配着羽毛灯，给人一种干净、清爽的感觉。

本案融入了现代时尚元素和传统元素，在设计上强化主题，注重突出文化内涵并挖掘餐饮文化的精髓，让消费者在视觉、味觉上火的全新认识和感受。

The sofa background of the private room is a huge monochrome hand drawing of the millennium town-Chongqing. Some rattan wallpapers with a lot battens hang on the surface of wall. Chinese-style decorated cabinets with aging treatment and wrought iron droplight embody cultural characters of aged Chongqing city.

The public washroom continues elements of corridor ground. The wall and the ground of the whole washroom are paved on little pattern tiles. The red cabinets attract attention extremely. The decoration that Jazz white marbles match with feather lamp set in private washroom makes people feel clean and comfortable.

This project applies modern fashionable and traditional elements, highlights the design theme, and cultural connotation, shows up the soul about culture of food and cooking, and makes people get a completely new understanding and feeling in terms of sight and taste.

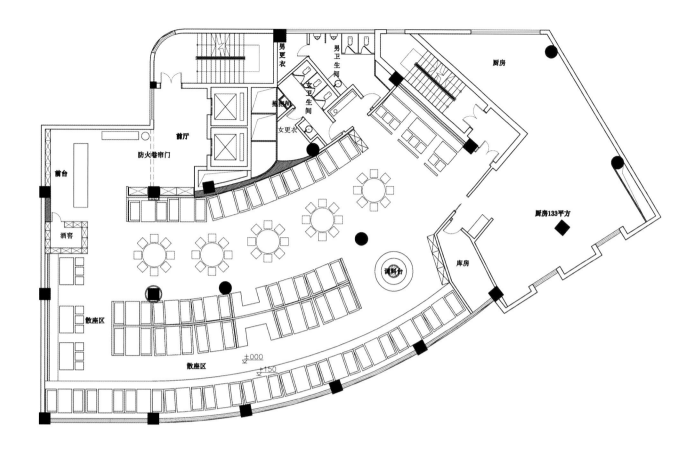

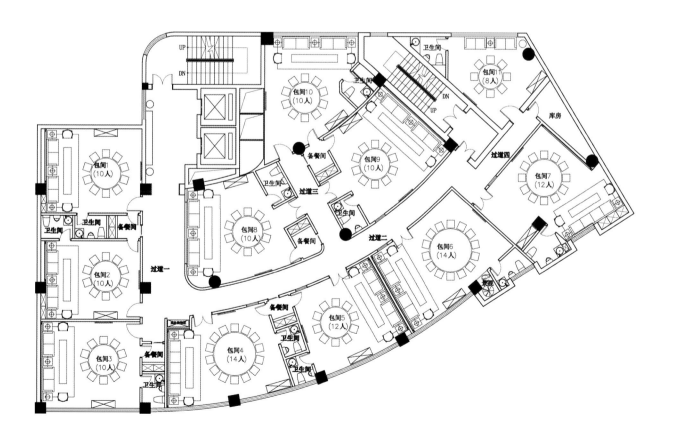

大厅的柱子采用灰白相间的马赛克，墙面是千年古镇重庆缩影的黑白手绘画，体现了重庆本土的建筑文化特色。彩色玻璃与黑色方管组合而成的隔断，既通透又能形成独立的空间，使客人在用餐时互不干扰。

彩色的玻璃、桌椅给素色的空间增加了许多色彩，也活跃了用餐氛围，给人一种热闹沸腾的感觉。原顶保留了原有空间的高度，裸露的管道更增添了空间的历史感。地面的桌椅在灯光的照射下反射到顶面镜面不锈钢的桥架上，形成一种倒置的感觉。

工业风格的黑色铁艺小吊灯似曾相识，有一种小时候家里吊灯的味道，与周围的环境相得益彰。靠窗的散座区做了一个抬高，很好地划分了空间的区域。镜面马赛克装饰的调味台显得内凹的空间更为宽大，黑色的台面上摆放着白色的陶罐内插着金黄色的麦穗，在灯光的映衬下，显得更为精致、美观。

The lobby pillar is laid on gray and white mosaics, the black and white hand drawings of the millennium town-Chongqing, hang on the wall, which embodies local architectural character of Chongqing. The grillwork combined with colorized glasses and the black square tubes, is transparent and forms independent space, which let customers have dinner quietly.

Colorized glasses and desks color the plain space and animate dinner atmosphere, creating a lively atmosphere. The ceiling keeps original level, and exposed tubes have a sense of history. In the light, the top stainless steel bridge reflects upside down shadow of desks on the ground.

The black little industrial-style droplight seems to have met before, like that in childhood, existing with the surroundings in harmony. On the side of window, the relax zone is lifted for an isolated area. The plate for condiments is decorated with mirror mosaics, which expands the concave space in visual. A white stean with a few golden wheats in it looks more delicate and beautiful under the light.

璞玉展厅
Puyu Exhibition Room

项目名称：璞玉展厅
项目地点：新疆乌鲁木齐
建筑面积：360 m²
主要材料：黑色花岗岩、白砂岩、竹帘
主案设计：蒋国兴
空间摄影：牧马山庄空间摄影机构 吴辉

Project name: Puyu Exhibition Room
Project location: Urumqi, Xinjiang
Building area: 360 m²
Main materials: Black granite, white sandstone, bamboo curtain
Project designer: Jiang Guoxing
Photograph: Wu Hui (Graze horse villa photograph organization)

　　玉石是中国传统文化的一个重要组成部分，以玉为中心载体的玉文化，影响了古代中国人的思想观念，成为中国文化不可缺少的一部分。
　　本案用现代中式风格展示玉器，在平面布置上，规划了进厅、大厅、展示区、洽谈区、办公区等，功能合理，动线流畅。在色彩运用上，以白色、木本色为主色调，黑色为辅色调，营造了一个素雅、别致的环境。

Jade is an important part of the Chinese traditional culture, the jade culture taken jade as central carrier affects conception of ancient Chinese and becomes indispensable part of Chinese culture.

The project displays Chinese jade with the help of modem Chinese style in the plane layout. It designs hallway, lobby, exhibition area, office and so on. The function is reasonable, and the routine is fluent. White and the color of log is the major tone and black is minor in the color application, which create a plain and unique environment.

进入进厅,对面是白砂岩LOGO墙面,不施任何粉黛,素颜装饰的墙面显得干净利落。白色竹子装饰的干景,在射灯的照射下,散发出一道道光影。左边是通往二楼的楼梯,悬空的设计使空间立刻变得灵动起来,楼梯下的白色细砂形成一圈圈水波纹的造型,几个精美的装饰品静静地立在那里。在进厅的尽头,墙面上有一个镂空的圆形,使两个空间之间有了简单的交流。

大厅没有复杂的造型,一条长长的大桌面立在中央,枯枝装饰着整个台面。墙面的柜子上摆满了各种各样的玉器,在白色灯膜的衬托下,愈发显得耀眼。在大厅的另一面墙面上,设计了一个电子显示屏,时时刻刻播放着最新的资讯。洽谈室与大厅没有完全隔开,而是延续了进厅圆的造型,两边是透明的玻璃门,既保证了私密的需求,又保证了空间的通透性。大厅的另一边做了一个地台,柱子将整个地台分成了两个洽谈区。木本色的家具、竹子的卷帘、黑白的挂画、黑绿搭配的抱枕、佛像,构成了一幅中式风格的画面。

Entering into the hallway, the logo wall across from white sandstone is plain and clean. The dry landscape decorated with white bamboos shows shadow under the spotlight. The stair leading to the second floor is designed to overhead, vivifying the space immediately. Under the stair, white sand form some shapes like water waves, and a few ornaments set on it. At the end of hallway, there is a hollow circle on the wall, which connects two spaces simply.

The design of lobby is concise, a long and big desk is set in the center, and some dry branches decorate it. Cabinet on the wall is set all kinds of jades, which look more sparkling in the shiny white screen. There an LED screen on the other side of wall in the lobby, the screen shows the latest news at moments. Reception is half-connected with lobby, and continues the circle shape of lobby. The transparent glass doors on the both sides, guarantee privacy and space connection. There is a platform on the other side of lobby, pillars divide the platform into two parts for negotiation. Log furniture, bamboo curtains, black and white paintings, black and green pillows match perfectly like a beautiful Chinese painting.

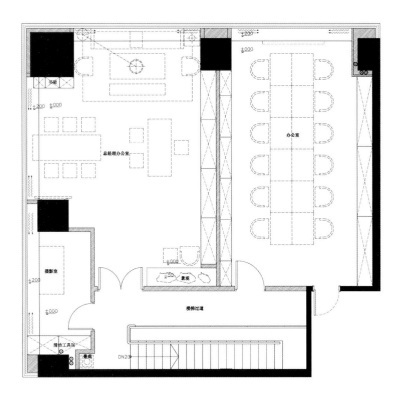
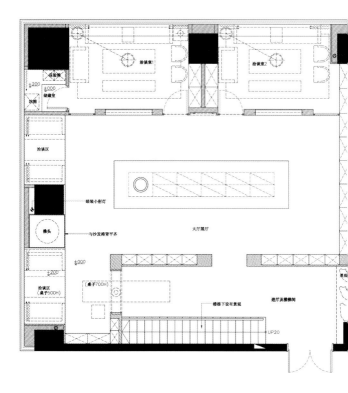

二楼主要以办公空间为主，分为董事长办公室和办公区。白色方管与实木层板组合而成的书柜摆满了各种各样的书籍，木作的上下柜提供了很充足的储物空间，中间墙面的清镜，在视觉上拉伸了空间感。裸露的原顶保留了原有空间的高度，一排黑色的吊灯整齐地立在办公桌上方，发出淡淡的光，强化了空间的柔和度。

董事长办公室门口的枯竹景观点缀着整个空间，墙面的木作柜子摆满了各种装饰品，粗犷的圆木整齐地堆放在书柜中，形成一道别致的风景。墙面采用的是白色木线条和藤制壁纸，与沙发的坐垫、靠枕相互呼应。

The second floor is office area, chairman office and staff office area, bookcase made of white tube and wooden plates is put in various books, wooden cabinets provide adequate space for storage, and mirrors on the middle wall expand space in visual. Exposed ceiling keeps the original level, an array of the black droplight sets over the office desk with soft light, making space soft.

Dry bamboo at the entrance of chairman office is an attractive decoration. The wooden cabinets on the wall are put various ornaments and rough logs are put in a bookcase in order. The white batten and rattan wallpapers decorate the wall which match with cushions and pillows on the sofa.

靓汤房子·泰式海鲜火锅
Delicious Soup House, Thailand Seafood Chafing Dish

项目名称：靓汤房子·泰式海鲜火锅
项目地点：新疆乌鲁木齐
建筑面积：430 ㎡
主要材料：灰色大理石、深色实木复合地砖、木拼条、鹅卵石
主案设计：蒋国兴
空间摄影：吴辉（牧马山庄空间摄影机构）

Project name: Delicious Soup House, Thailand Seafood Chafing Dish
Project location: Urumqi, Xinjiang
Building area: 430 m²
Major materials: Gray marble, dark wood compound tile, batten, cobble
Project designer: Jiang Guoxing
Photograph: Wu Hui (Graze horse villa photograph organization)

　　本案的定位为大众化的百姓消费场所，同时体现一定的文化主题，为火锅的经营赋予更有层次的文化内涵。因此在环境和空间设计上充分考虑其独特性和新颖性，体现餐厅的特点和优势以吸引更多的消费者。
　　空间基本功能划分为：等候区、卡座区、包间等。在色彩运用上，以原木色、土黄色为主，搭配黑色的吊灯、白色的挂画、绿植，营造了一个新颖、别致、时尚的用餐环境。

The theme of project is popular civilian consume, and it embodies a kind of connotation of chafing dish culture, the design of whole atmosphere and space is unique and creative, showing restaurant's characters and advantages for attracting more customers.

The basic functions of the space include: waiting area, deck area, private room, etc. In the application of color, the color of log and earth is primary tone, it creates a sense of unique and fashionable matched with black droplight, and white paintings and green plants.

缓缓走上二楼，墙面挂着一幅别致的画，没有任何色彩，几根枯木树枝组合而成，好像经历过岁月的洗礼，在灯光的照射下愈发显得别致，与顶面白色的枯木树枝相互呼应着。黑色方管与实木板组合的栏杆与等候区的装饰架融为一体。等候区的墙面采用原木色的木拼条，木质的层板上摆放着一些装饰品和一盆绿植，绿藤很自然地垂下来，给素色的空间增添了一点活力。等候区还设计了一个内凹的电视，可以播放着最新的菜品或者当下最热门的话题，使人在等候的时候也不会觉得无聊，人性化的设计更提升了餐厅的服务品质。

Get to the second floor, you can see a delicate painting without any color on the wall which is made of some dry branches, like spending a long time. It becomes more delicate in light, matched with the top white branches. Railings combined with planks and black square tubes are well matched. Log batten on the wall in waiting area, some ornaments and plants are set on the wooden lamination, the green rattan hangs naturally, all of which vivify the space. There is a concave television in waiting area, which can screen the latest news and the hottest topic, you can never feel boring during the waiting time. The humanized design also improves the quality of the dining room.

整个大厅功能布局合理，动线流畅。服务台摈弃了常规的大理石设计，而是以实木装饰台面，鹅卵石装饰侧面，再外置一层铁丝网，别致又独特。台面上面是黑色方管和玻璃定制的吊架，摆放着大小不一的陶罐和绿植，绿藤缓缓地垂下来。卡座与卡座之间都是以矮隔断隔开，既很好地保持了视线的穿透性，又很好地保证了用餐环境，使客人在用餐时互不打扰。墙面挂着大小不一的手绘画，增添了更多的原创性。为了融入大厅的环境，设计师在原有窗户的基础上又做了一层造型，黑色方管与铁丝网相结合，搭配黑色百叶帘，既美观又可以使阳光照射进来。大厅原顶保留了原有的高度，而裸露的管道给人一种工业风的感觉。

The layout of lobby is rational, which fluent routine. The service counter's surface material is wood rather than marble, and the side is decorated with cobble, covered with a layer of iron net, which is unique and attractive. Over the desk, there are black tubes and glass cradles which are set in steans and hanging green plants. A grillwork put between two lounges guarantees quiet dinner environment and visual penetrability. All sizes of paintings hung on the wall intensify the originality. For harmonizing the environment of lobby, the designer designs another shape on the basis of original windows. Black tubes matched with iron nets and black louver, is beautiful and can let sunlight in. The ceiling of the lobby keeps at the original level, and exposed tubes are full of industrial style.

包间没有复杂的造型，鹅卵石、木拼条、条砖装饰着整个墙面，再搭配一些挂画、装饰架、铁艺灯具，处处散发出一种时尚又别致的气息。

卫生间延续了大厅的风格，木拼条装饰着整个墙面，卫生间的隔断也采用了木拼条，与墙面融为一体。

The design of private room is concise, using cobbles, battens and bricks to decorate the whole wall, which matches with some paintings, ornamental shelves and iron lights, showing a kind of fashionable and unique sense.

Washroom has the same style of lobby: battens are used on the wall and the partition of the washroom.

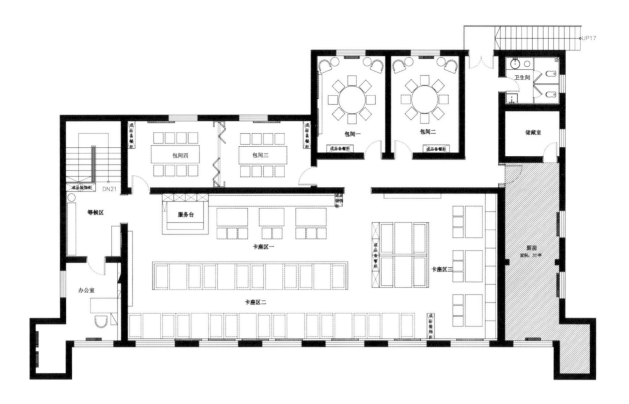

美亚巨幕电影院
Meiya IMAX Cinema

项目名称：美亚巨幕电影院
项目地点：新疆乌鲁木齐
建筑面积：5000 ㎡
主要材料：中国黑荔枝面环氧树脂、自流平、黑钛金不锈钢、灰镜
主案设计：蒋国兴
空间摄影：吴辉（牧马山庄空间摄影机构）

Project name: Meiya IMAX Cinema
Project location: Urumqi, Xinjiang
Building area: 5000 m²
Main materials: Epoxy resin of black bush-hammered facing, self-leveling, black titanium stainless steel, gray mirror
Project designer: Jiang Guoxing
Photograph: Wu Hui (Graze horse villa photograph organization)

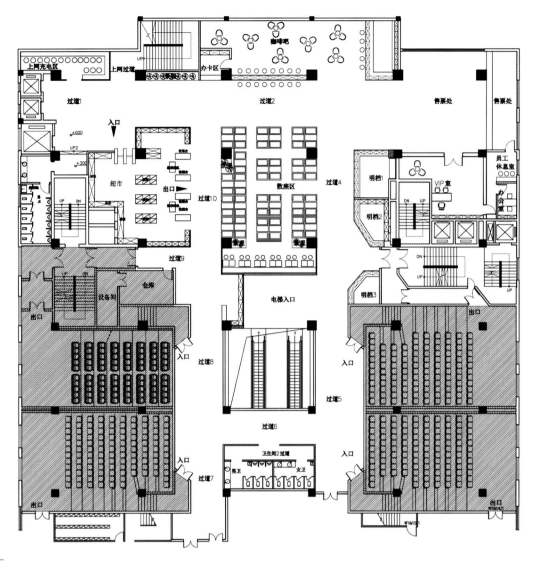

3F平面设计图

美亚巨幕影城位于新疆乌鲁木齐，聚集于娱乐、餐饮为一体的商业圈，高端的市场定位，很好地诠释了未来城市生活空间的发展远景。

为了体现文化的历史特性，本案不会追求过度的娱乐包装、盲目地搭配造型、令人炫目的怪异灯光，而是谋求整体上的简洁大气、和谐统一，但又不失时尚和新潮。颜色的搭配、空间形体的塑造、材料的选购、施工工艺的选用都体现了整个空间的市场定位。

Meiya IMAX cinema locates in a commercial area for amusement and catering in Urumqi, Xinjiang. Its market positioning is advanced consumption, and makes a good explanation to the future development of city life space.

For embodying the historical characters of culture, the design of the project pursues Integral concise and fashionable sense instead of excessive amusement package, irrational shape, and monstrous dazzling light. The color scheme, the shape of space configuration, materials and construction technology reflect market position.

售票区以一幅巨大的电子显示屏为背景，时时刻刻为来往的客人播放最新的电影资讯。一整排的黑金沙大理石台面，简洁又大气；做旧的耐候钢装饰的服务台立面，与台面形成鲜明的对比，又好像历经沧桑，像生了锈的铁一样静静地立在那样，默默地注视着每一个过往客人。售票区旁边放着一组黑色的沙发，为客人提供了一个舒适的等候区。

咖啡区的顶面和地面均采用做旧的深色木板，两者相互呼应着。黑色方管与实木层板组合的酒柜摆满了各种各样的咖啡及饮品，满足客人不同的口味需求。实木层板与耐候钢组合的吧台，延续了之前的元素与风格。吧台上方的黑色吊架，摆放着一些土陶罐，很好地装饰了整个空间。

每个影厅门口的旁边都做了一个独特的LOGO，耐候钢雕刻数字的造型内置灯管，通过透光板发出淡黄的灯光，既满足了装饰的需求，又很好地起到了引导的功能。

四楼大厅中国黑荔枝面环氧树脂地面，复古红砖与做旧的木拼条装饰的墙面，裸露的原顶以及黑色的桥架，无一不透露着工业风的情怀。

The background of ticket office is a giant LED screen which plays the latest news about movie information; an array of black Galaxy marble table-board is concise. Retro climate-resistant steel covers facade of service counter, which contracts with surface intensely. It also looks like a rusty iron, set on there in silence, like a witness of nowadays and yesterday. Beside the ticket office, there is a set of black sofa, and guests can have a rest there.

The ornament material of the ceiling in coffee area is retro dark floor, the same as the ground. There are all kinds of coffee and other drinks for people's different tastes on the cabinet made of black tubes and wooden laminate. The bar counter combined with wooden laminates and climate-resistant steels has the similar element and style as ticket office. Some steans are put in black hanging shelves over the counter, which is an attractive ornament in the whole space.

Beside each door of the cinema ball, there is a unique LOGO, which is made of climate-resistant steel and transparent board, and it emits yellowish light by inset light. It's good for guiding and beautiful.

The material of the ground in the lobby on the fourth floor is epoxy resin of black bush-hammered facing, retro bricks and aged-like batten decorate walls, together with exposed original ceiling and black bridge, revealing a sense of industry-style.

057

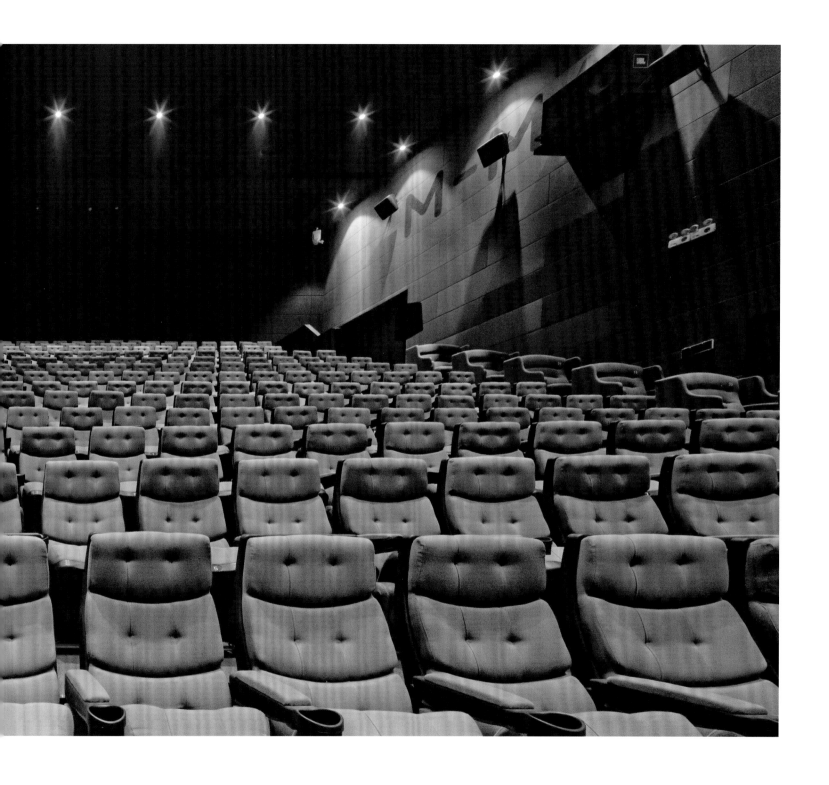

　　两层的空间结构布局合理、动线流畅。在平面布置上规划了上网充电区、咖啡区、美食区、售票区等。人性化的设计，告别了传统单一的影院模式，提升了影城的服务品质。

　　三楼大厅咖啡色的自流平地面，颜色深浅不一，自然流畅。墙面装饰着复古的红色条砖、做旧的素色水泥墙面、做旧的木拼条，质朴粗犷，仿佛带着泥土的气息，亲切自然，给人一种怀旧的感觉。黑色方管组合的架子、黑色铁丝网隔断、黑色桥架、顶面裸露的管道时尚又前卫，仿佛让我们看到了后工业时代的缩影，与墙面质朴粗犷的材质和谐又统一。

　　The layout of the two-level structure is rational, motion routine is fluent. There are areas for internet, coffee, dinner and ticket-selling in plan layout. The humanized design instead of usual and monotonous cinema pattern, promotes the service quality.

　　The self-leveling ground in the lobby on the third floor is mixed brown and looks fluent. The wall is decorated with retro red bricks and plain cement, aged-like battens are rough with a flavor of natural sense, creating a nostalgic sense. Shelves made of black tubes, black grillwork of iron nets, black bridge and exposed tube over the head are fashionable and have a sense in the post-industrial era, matching with rough and plain materials in harmony.

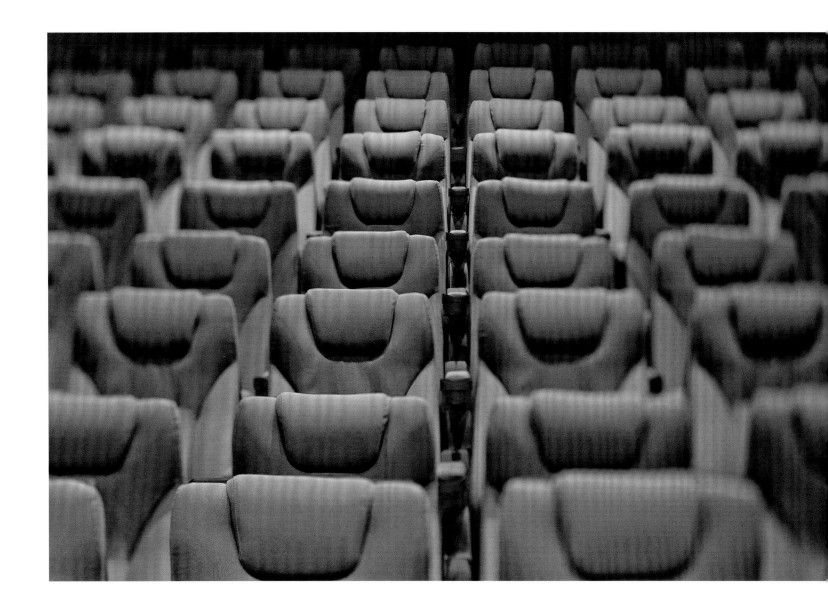

令人难以忘怀的是大厅的前厅，木拼条装饰的墙顶面，W形的造型，好像一条长长的回廊，仿佛回到了以前台北车站的小资情调，让人久久不能忘怀。

前厅的对面规划了一个别出心裁的景观，玻璃隔断里面摆满了各种各样的乐器，时而高、时而低、时而直立、时而倾斜，丰富的姿态好像有人在里面任意地抒发着情绪，给偌大的空间增添了一首无声的音乐。

影厅的走道没有复杂的造型，复古的红色条砖铺满了整个墙面，在走道的中央规划了几个弧形的哑口，白色的小灯装饰着弧形的哑口，发出一圈淡淡的灯光，反射在地面，形成强烈的对称视觉感官，装饰着整个过道，在视觉上缩短了走道的长度。

卫生间采用了黑、白、灰的色调，黑色的瓷砖、做旧的水泥墙、黑钛金边框、黑色烤漆玻璃，给人一种冷色调的硬朗感。

Lobby's anteroom is impressed, the wall and ceiling of which are decorated with batten. W-shaped form is like a long corridor at railway station in Taibei before.

There is a unique landscape across from the lobby, where many instruments are set in glass grillwork. These instruments are set in high or low level; ones are hang straightly, the others are set on the tilt; the rich gestures seem that there is someone within it express his emotion, adding silent music to this large space.

The corridor of cinema hall is concise, retro and red bricks adorn the wall. A few arcs passing in the center with white lights reflected on the ground, it shortens length of corridor in visual.

Washroom takes black, white and gray as the main tone. Black tile, retro cement wall, black titanium frames and black stoving varnish glass reveal tough sense of cold tone.

深海壹号
No.1 Deep Sea

项目名称：深海壹号
项目地点：新疆乌鲁木齐
建筑面积：2080 ㎡
设计公司：叙品设计
主要材料：深色实木地板、灰色花岗岩、深蓝色砖、蓝色壁纸
主案设计：蒋国兴
空间摄影：吴辉（牧马山庄空间摄影机构）

Project name: No.1 Deep Sea
Project location: Urumqi, Xinjiang
Building area: 2080 m²
Design company: Xupin Design
Main materials: Dark floor, gray moonstone, mazarine tile, blue wallpaper
Project designer: Jiang Guoxing
Photograph: Wu Hui (Graze horse villa photograph organization)

现代人们在品尝美味佳肴的时候，开始关注用餐环境的文化氛围和个性化，而餐厅内的设计正是为满足顾客的这部分需求。在这个像大海一样不可测度的魅力餐厅内，充满宁静、神秘、新奇，在你身临其中的时候，希望通过视听与联想，能让你进入期望的主题情境。

Nowadays, people start to focus on cultural atmosphere and personality of dinner environment. The design of dinner room is to satisfy this necessary. This ocean-themed dinner room which is like the sea, is filled with mysterious and charming, peaceful and unique breath, you can get into expected theme situation when you are here by the way of seeing, listening and imagination.

深海壹号位于新疆乌鲁木齐南湖东路酒吧园区，是近年新崛起的商业区，聚集于娱乐、餐饮为一体的商业圈，主营海鲜火锅，是一家以海洋为主题的高端餐厅。

设计师与业主沟通后，对海洋概念的了解更深一步，形成一个用概念手法诠释海洋文化的主题餐厅。

不同的餐饮文化主题会给人不同的感受，以海洋为主题的餐厅呈现出一种蓝色的氛围。蓝色调体现着丰富的情感内涵，蓝色让人联想到广阔、深远、蓝蓝的天空，波浪滔滔的大海，忧郁的情感也总是与蓝色连在一起，使人感到优雅、宁静。金色是餐厅的第二个色调，主要分布在餐厅顶部，顶部保留原结构造型，增加线条装饰，配以暖色灯光，营造明亮、华丽、辉煌的视觉效果。金色宛如轻柔的沙滩，沙滩与海相辅相成，海浪静悄悄地通过，又悄悄地退去，聆听着海风，感受着沙滩的柔软、温暖，感受太阳的气息，享受着美好时光。

进入大厅入口，首先映入眼帘的是一只做旧的大型古帆船。经过帆船你会观赏到船头礁石上美人鱼的动人姿态。大厅布局上动线流畅，没有阻挡，而地面则采用各色条形大理石铺设，增加了大厅的层次感，犹如海浪让人觉得自己不是在走路，而是驾着一艘舰艇穿行在辽阔无际的海洋上，自由而又惬意。帆船的上方，悬挂成群的银鳞小鱼，如有生命的鱼儿嬉戏追逐，成群结队，让人更加亲近自然、热爱生活。海浪、沙滩、鱼群、古船、鸣笛、深海生物造型的纸灯，还有老船长，这些都不禁勾起对童年时的回忆。大厅右侧是服务台区，服务台前部由白炽灯泡装饰，配上蓝色灯光，流光溢彩，有种梦幻质感。左侧设置了一个等候区，圆形的卡座沙发融入环境，休闲舒适，用餐前客人可以在这里小憩一会儿，可以与友人聊聊天，也可以欣赏餐厅主人安排的音乐节目。再往两侧去是餐厅的散座区和开放散包，散座以鱼鳞状隔断分隔，蓝色的麻布沙发，充满海洋的气息。而开放散包用灯泡装饰隔墙，宛如深海的泡沫向上漂浮，配上蓝色灯光营造的氛围，浪漫迷人。蓝白的椅子增添空间的活跃感，再加上精心挑选的餐具，共同营造出高尚、雅致的用餐环境。

No. 1 Deep Sea locates at Bar Garden Area, Nanhu East Road, Urumqi, Xinjiang, which is a new commercial and entertainment district. It is an ocean-themed high-end restaurant, specialized with seafood hot pot.

After communicating with the owner, the designer gets a deeper understanding of the ocean and creates a theme restaurant of marine culture by a conceptual approach.

Different themes of catering culture give different feelings. Ocean is blue and deep. Blue is always associated with sky and sea, and it is also used to represent elegance and peace. Gold is the minor color used on the ceiling, which keeps original shape and adds line decorations, together with warm light, creating a bright and elegant effect. The combination of gold "beach" and blue "sea" is wonderful and enjoyable: silent waves, sea breeze, soft beach and warm sunlight.

Entering the lobby, you will see a huge old boat, passing through which a mermaid on the rocks is in sight. The motion routine of the lobby is reasonable and smooth. The ground is paved with colorful marbles, which looks rich in layers. On the right side of the lobby, there is service area. On the left side is the waiting area. Round sofas are matched with the environment. Going through the lobby, there is dining area. The tables are separated with partitions of fish scales wood, together with blue linen sofas, forming a romantic sea atmosphere.

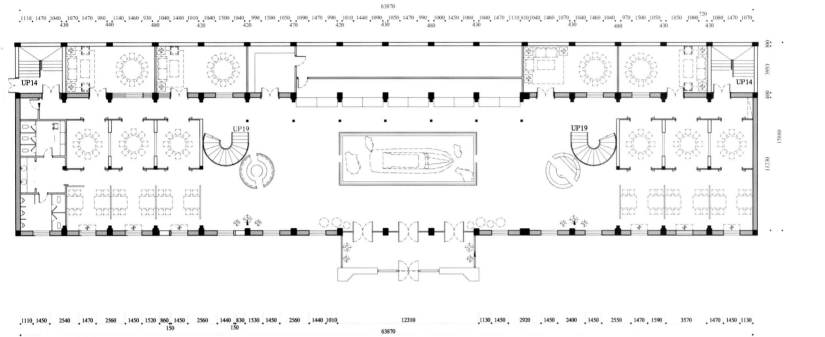
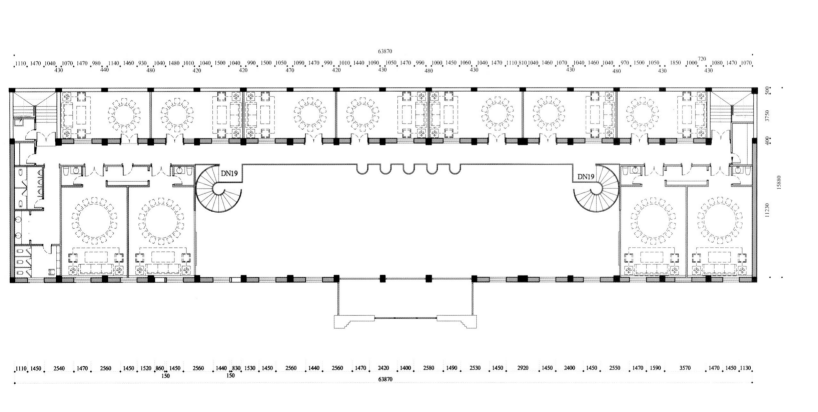

包厢部分集中在餐厅二楼，大包厢的顶部延续了大厅的元素，保留原结构以金箔贴顶加以石膏线条，蓝色的壁纸墙面，突显海洋主题，并将活泼好动的海洋生物饰品融入环境当中，在安静的空间内增添了趣味性。包厢并不是全封闭的，在包厢靠大厅的侧面墙，设计师运用鱼鳞隔断预留位置，在来客用餐的同时也可以欣赏大厅的音乐节目，丰富用餐乐趣，把周围的环境应用并调动起来，充满人文气息的知性风格，融入了人们对生活的品味和期许，优雅、舒适。

The private rooms are focused on the second floor of the dinner room, the top decoration of large private room continue lobby's elements, and keeps original structure, and it is pasted on gold foils, gypsum line around the edge of ceiling, blue wallpaper and ocean biology ornaments show the ocean theme, adding interesting characteristics to the silent space interesting. The private room are half-open, the side wall of room near the lobby, designer applies scale-shaped partition reserve a place for guests to appreciate music program during dinner time, enrich the dinning pleasure. By applying and arousing the surrounding environment, perceptual style with literary sense melts in people's taste and expectation about life, which is elegant and comfortable.

醉玥餐厅
Drunk Yue Restaurant

项目名称：醉玥餐厅
项目地点：新疆乌鲁木齐
建筑面积：1500 ㎡
设计公司：叙品设计
主要材料：镜面不锈钢、黑白红高亮砖、铁丝网、做旧条砖
主案设计：蒋国兴
空间摄影：吴辉（牧马山庄空间摄影机构）

Project name: Drunk Yue Restaurant
Project location: Urumqi, Xinjiang
Building area: 1500 m²
Design company: Xupin Design
Main materials: Mirror stainless steel, black and white red bright brick, iron net, retro brick
Project designer: Jiang Guoxing
Photograph: Wu Hui (Graze horse villa photograph organization)

本案的顶面立面造型不作过多复杂的设计，只进行简单的处理，在色系上以中间色为主，地面的红色活跃氛围，通过灯光营造与材质的运用，来丰富空间。现代时尚、浪漫复古的醉玥餐厅，应是年轻人所喜爱的。

The design of the top and facade is concise, the color tune is neutral color, the red color on the ground vivifies the atmosphere, and the application of light and material enrich the space. I believe that the modern and romantic retro restaurant is popular among the youth.

本案是新疆乌鲁木齐东街项目综合体的一个对外中餐厅，是一个主营中餐的中高端餐厅，主流客源以年轻人为主，所以设计师考虑以年轻化作为设计主线；以现代欧式，并略加当下流行的工业风来渲染空间氛围。以做旧的材质来演绎复古怀旧情怀，现代与怀旧的结合，营造出散发浪漫气息的时尚餐厅，来引领与提升年轻人的消费方式与生活品质。

The project is an integrated Chinese restaurant located on East street in Urumqi, Xinjiang province. It is a superior restaurant that specializes in Chinese-food, and faces to young people. The designer makes space sense of young and alive by the elements of modern European style with some popular industrial style. Such materials with antique finish make people return to the ancient. Modern and retro feeling filled in this romantic and fashionable restaurant leads young people to advance the way of consumption and the quality of life.

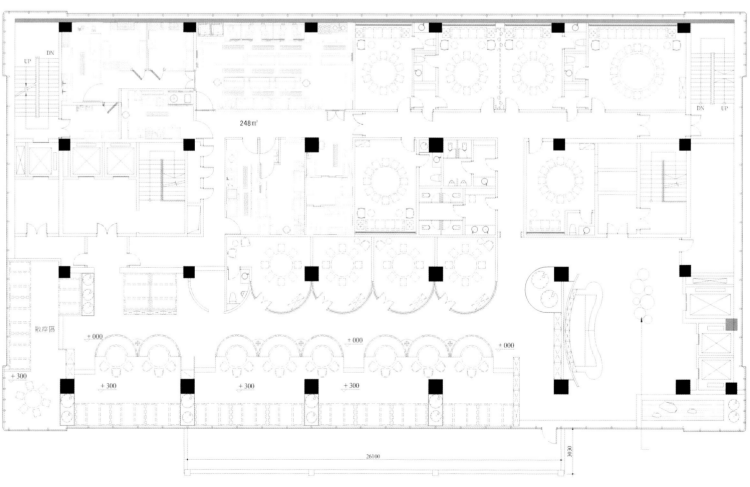

在平面的规划上，功能区域划分分明，主入口直对餐厅的服务台，便于来客直接享受店内服务，更利于服务员的快捷引导。服务台运用镜面不锈钢材质，将地面黑白红马赛克砖映射得更加丰富，设计师特意在黑白马赛克砖穿插点缀红色，更加活跃了空间氛围，在顶面异形球灯的时尚造型下，灯光增加了空间的温馨感，使空间中漂浮着一丝迷离的味道。前厅的中空地带，人性化地设置了不锈钢定制的不规则圆形矮凳，以供来客等候之用，在闲暇之余，翻阅手边的杂志书籍。经过前厅服务台，转角遇见一处芭蕉绿植景观，一片绿色，消除疲倦。芭蕉是热情的象征，体现店家的迎宾之道，再往前走是开敞的散座空间，设计师将餐厅最后的位置留给了来客，沿窗的落地玻璃，敞开的视野，享受午后阳光的扬洒，夜晚星空的迷醉，让客人心情舒畅，勾起别样的情怀。散座的中间，设计一排圆形卡座，丰富的造型，给来客更多可选性，也打破了空间的局限性，与对面的圆形卡座相互呼应。圆形卡座背面同样运用镜面不锈钢，延续材质的手法在视野上增加空间感展。半圆形包厢外用铁丝网分隔，半通透的铁丝网消除包间的局限感，就餐于包间的来客，隐隐约约间可以透过落地窗欣赏外面蔚蓝的天空。回头再经过过道，过道的另一区域是餐厅的包厢区，与散座区分隔两边，提供了安静的就餐环境，也提升了就餐时的私密性，包厢的氛围与外围极具现代感，温馨怀旧的情怀直入人心以及墙面的透光假窗给予包厢通透性，不规则的黑白挂画叙说过往，做旧处理的木地板，做旧的木假梁以及墙面的做旧条砖，随性而感性，在鹿角灯的烘托下，仿佛在讲述一个故事。家具在整个环境里也是重要的一环，经过设计师精挑细选的欧式家具，散落在餐厅中的每一处，它们也是故事中的一员，柔化了空间。

In fixture and furniture plan, designer divided function area rationally, main entrance across from service counter is convenient for serving guests. Surface of the counter is made of mirror stainless steel, reflecting black, white and red mosaics on the ground. The color of red is designed in black and white by designer on purpose to vivify the atmosphere. The overhead irregular-shaped spherical lamp emits warm light in fashion space that make a sense of mysterious. In the center of anteroom, there sets many irregular circle short stools supply for waiting. In leisure time, you can read magazines for pastime. Go by service counter of anteroom, you can see a banana plant landscape at corner. The color of green remove exhaustion, and the banana tree is the symbol of passion, like owner's personality to guest. Go straight to the road, it's an open lounge area. Designer leave the final location to guest, when you stand at the side of French glass on the edge of window, you will see a vast scene, feel the sunshine in the afternoon, and see the amazing sky at night, and you will feel comfortable, and recall unique memories. In the center of scattered sites, there is an array of circle lounge, many kinds of sculptures give guests more choices and break the boundary of space style. It is matched with the opposite circle soft roll. On the back of circle lounge, the material also applies mirror stainless steel, which expands the space in harmony. Semi-circular private room is divided with iron net. Translucence iron nets break the boundary of the room; guests who have a dinner in private room can see the blue sky through French window indistinctly. Return and go by the corridor, the other district of corridor is private area of dinner room which set on both side of leisure area. It supplies a quiet dinner environment and protects privacy. The warm and friendly atmosphere of private room is different absolutely with the modern and industrial style outside. The transparent fake window on wall lights the room, irregular black and white painting is like an aged story. The retro floor and wooden fake girder and bricks of wall are all casual and emotional. In the antler light, a story starts. Furniture is an important part of the whole environment, carefully chosen furniture of European-style set on the ground of dinner room.

新空气健康管理中心
New Air Health Management Center

项目名称：新空气健康管理中心	Project name: New Air Health Management Center
项目地点：新疆乌鲁木齐	Project location: Urumqi, Xinjiang
建筑面积：4200 m²	Building area: 4200 m²
主要材料：白色人造石、藤编壁纸、海藻泥、木饰面等	Main materials: White artificial stone, rattan wallpaper, algae clay, wooden veneer
主案设计：蒋国兴	Project designer: Jiang Guoxing
空间摄影：吴辉（牧马山庄空间摄影机构）	Photograph: Wu Hui (Graze horse villa photograph organization)

提到医院，我们马上会联想到白色，白色的墙面、地面、白色的制服，给人一种严肃的感觉。本案以关注人体健康为出发点，引入禅学环境，颠覆了传统的医院模式。

在平面布置上分为两层，整个空间运用了暖色调的米色系，与原木色相结合，给人一种家的感觉，使人不再感觉恐惧。前厅以沙漠、草原、绿植等元素为背影，一进入大厅便能感受到一种空旷、寂静的回归之美。服务台摈弃了常规的大理石做法，以黑色方管和木头相结合，质朴中透露着亲切的感觉。工业感的吊灯与空旷的背景形成鲜明的对比，又与黑色的方管很好地呼应着。接待区放置米色的沙发，搭配木质的小桌椅、白色的层板架、金黄色的麦穗，简洁又不失温情。装饰壁炉无疑是整个接待区的点睛之笔，整齐排列的小木头、黄色的火焰温暖着每一个等待的病人，此刻或许已忘记了自己是个病人，好像是坐在咖啡厅，喝着咖啡，翻看着杂志，享受着生活。

Speak of hospital, we immediately think of the color of white, white wall and ground, white uniform, all of which give us a serious feeling. The project overturns traditional pattern. In order to focus on health, we introduce Zen environment, and create a kind of popular industry in future.

There are two floors in plan layout. It gives people feeling of at home by applying warm yellowish color, matched with log color, the fear is gone. The background of anteroom apply elements of desert, prairie and plants, you can feel spacious and silent as entering the lobby. Service counter is decorated with black tubes and wooden laminate instead of usual marble. It makes a sense of warm and friendly rather than cruel sense. Industrial-style drop light and wood desks and chairs contrast intensively with spacious background and echo with black tubes. There is beige sofa matching with small wooden tables and chairs in reception area. White laminates and gold wheat are concise and tender. It's no doubt that ornamental fireplace is the most attractive point in reception area. An array of battens with yellow fire makes every patient feel kind during the waiting time. Probably at that moment, patients forget they are ill; they can just sit there, drink a cup of coffee, read a magazine and enjoy life.

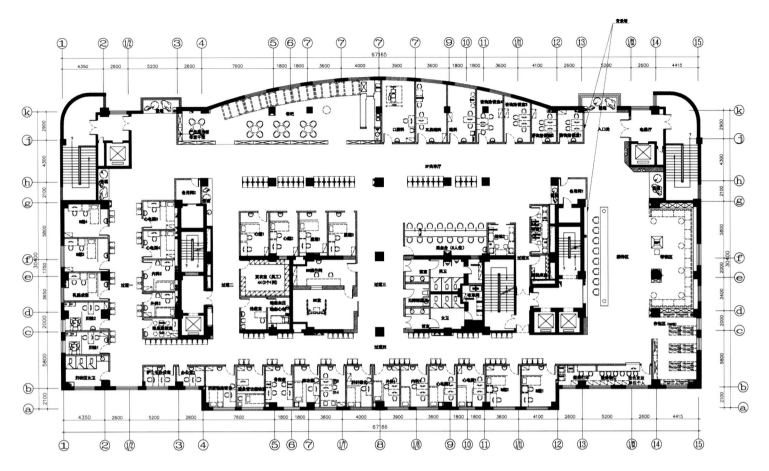

走道没有复杂的造型，白色的实木线条和浅灰色的壁纸装饰着整个三楼的墙面，颜色鲜艳的椅子给单调的走道增添了一些色彩。实木线条把四楼米色的壁纸墙面分成一格一格的，藤编的椅子与米色系的墙面很好地呼应着，挂画点缀着墙面。设计师在每一层还专门设计了一个餐吧区，人性化的设计提升了整个空间的品质。白色的层板架上摆放着大小不一的陶罐，还有各种各样的书籍。在这里人们可以慢慢地用餐，也可以悠闲地翻看着书籍，时间就不知不觉中过去了，不会觉得无聊。

There are no complex shapes in the corridor, white wood lines and light-gray wallpapers adorn the whole walls of the third floor, the chairs with bright color add some color to the dull corridor. The wall is divided into many lattices by white wood lines. Rattan chairs is matched with yellowish wall, paintings embellish the wall. The designer designs a dinner area on each floor, and the humanized design improve the whole quality of the space. All kinds of steans and various books are put on the white lamination. Guests can enjoy dinner there, or read magazines leisurely. The time passes away unconsciously and you never feel bored.

VIP室延续了整个空间的元素，质朴的材质装点着墙面，藤编壁纸、海藻泥、木拼条，木质的层板架、木质的沙发茶几，粗犷质朴中显出亲切的感觉。

三楼卫生间以黄白色为主色调，搭配白色的台面，看上去清新又自然。四楼卫生间以灰白色为主色调，灰色木纹大理石铺满整个墙面、地面，与灰色的木隔断相互呼应，白色的洗手台面设计了一个嵌入式的垃圾篓，看上去整洁又干净。

本案的设计朴实、自然、温情，设计理念为以人为本，不仅满足了功能的需求，更是一次愉快的体验，也将是健康产业包括医院环境体验的一次革新。

VIP chamber extends the whole space elements. some rustic materials are embellished on wall. The wall paper made of rattan, diatom ooze, batten, wooden storage rack and wooden sofa with tea table are make people feel simple, native and familiar.

The main tune of the washroom on the third floor is yellow and white, the white surface of desks is natural and fresh, the washroom on the fourth floor is gray and white, gray marble with wood grain are paved on the wall and the ground, matching with gray wooden partition. An embedded garbage can set in white wash basin looks clean and tidy.

The design of this project is plain, natural, friendly and people-oriented thought is the design concept. It satisfies necessary of function, and reform health industry including hospital environment experience.

苏州龙海建工
Longhai Construction Engineering Group, Suzhou

项目名称：苏州龙海建工
项目地点：江苏省昆山市
建筑面积：1800 m²
主要材料：深色木地板、浅灰色壁纸、灰色墙面砖、黑色木饰面
主案设计：蒋国兴
空间摄影：吴辉（牧马山庄空间摄影机构）

Project name: Longhai Construction Engineering Group, Suzhou
Project location: Kunshan, Jiangsu
Building area: 1800 m²
Main materials: Dark floor, light gray wallpaper, gray wall tile, black wooden veneer
Project designer: Jiang Guoxing
Photograph: Wu Hui (Graze horse villa photograph organization)

　　龙海建工办公室位于江苏省昆山市前进东路上的东安大厦。昆山是中国大陆经济实力最强的县级市之一，近几年的经济快速发展引来越来越多的外来企业入驻此地，龙海建工便是其中之一。
　　办公室是脑力劳动的集体空间，企业的创造性大都源自于该空间的个人创造性发挥，因此在这个场所内要注重个人的环境还要兼顾集体的空间，以活跃人们的思维，积极提高企业的工作效率，这也是企业提高生产效率的一部分。另外它也能体现企业的整体形象及文化内涵，更能提高客户对企业的信任感及员工心理上的满足感。

The office of Longhai Construction Engineering Group locates in Dongan Building, Qianjin East Road in Kunshan city, Jiangsu province. Kunshan is one of the most developed country-level cities in China. Because of the rapid economy development in recent years, many exotic enterprises are attracted here, and Longhai Construction Engineering Group is the one of them.

Office is the public space for mental workers. Creativity of a company is originated from each staff, the public environment is the same as important as private space. The active minds and attitude is one of ways to improve work efficiency. In addition, it can embody company's image and cultural connotations and it also can improve the staff confidence and psychological satisfaction.

本案设计师信守均衡对称的原则并以此来规划室内的布局，使得整个动线流畅明快，充分利用了各个空间，让空间布局更为合理；整个办公室以现代手法与设计师所理解的中式传统文化内涵及中式元素相融合，使整体庄重、淡雅。大厅的前台背景是特意在石材产地挑选的，它如墨如画般淡雅出尘，妙在似与不似之间再配以侧边的水景以及内敛、质朴的中式家具，让人觉得别有一番韵味，使人产生丰富的遐想，无形中透出中式的意境之美。过道用简洁硬朗的直线条勾勒层次感；以朴实的手法，通过虚与实、明与暗之间的交替使其显得简约而不简单。在过道拐弯处会发现一排树干修直、洁白雅致的白桦景观，它坚韧、挺拔、正直的精神是正企业文化的一种象征。

The project designer plans space layout by following the principle of balance. Route design is fluent and specific, all space are designed sufficiently and availably. All modern-style elements in the whole office are matched with all Chinese-style elements that are extracted from Chinese traditional cultures in designer's mind, that make a grave and elegant atmosphere. The background of lobby is stone which is chosen from stone producing area. The stone is plain and elegant like Chinese painting, the point is that its image is like or unlike something which makes a unique sense of Chinese-style art conception. The water scene on the side is matched with Chinese furniture. Corridor is a layering space which is decorated straight with lines. The alternation of abstract and concrete, light and shadow, makes a sense of concise instead of simple. When you go by the corridor, you can find an array of white and elegant birches with straight boles, its spirit of grittiness is a symbol of enterprise's culture.

办公室内部设置大面积的书架、摆放瓷器的饰品架以及东方主题的艺术陈设，这些书香气息陶冶着情操，同时彰显着办公室主人高品质的文化素养。关于顶部设计师则用简练的矩形浮面来体现，整体充斥着立体感，与灯光的碰撞让人感到视觉上的冲击，在这宁静、安逸的氛围中增添了一抹活跃，利于开阔思维。公司的中庭阳台处还有另一番景象，防腐木、墙面砖及自然的植物景观，人与自然的亲密接触，呼之欲出的休闲氛围，你可以在工作之余到这里来喝一杯咖啡，品一杯清茶，赏着夕阳与同事或者客户畅谈也不失为一件快事，让你从工作的疲倦中解脱出来。

There are many bookcases put in various books and many shelves put on chinaware in office, the orient-style ornaments reflect the owner's brilliant sentiment. The design of ceiling is concise rectangular floating face, the contrast of light and third dimension is intensive in vision, it can vivify the silent atmosphere and is good for open minds. There is another scene at balcony of atrium, it is a plant landscape combined with anticorrosive wood and wall tile. You can be close to the nature, breathe leisurely, drink a cup of coffee or tea, appreciate the sunset with colleagues or have a talk with clients, removing the exhaustion from work.

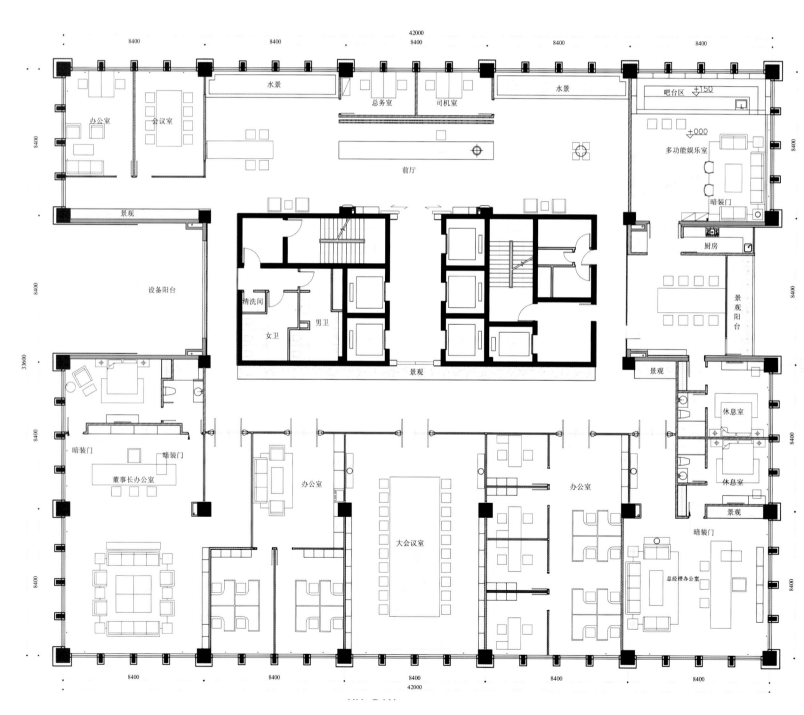

在现代烦琐的工作与快节奏的生活中提供一处宁静、淡雅、休闲的工作环境，在这里无论是员工还是客户都会得到自己心里所寻求的那份惬意。

Nowadays, the life and work are rapidly changing. This project provides a peaceful, elegant and relaxing environment for working. I believe that the office is an ideal place for clients and staffs to find the pleasure of their own.

叙品设计苏州公司
Xupin Design Suzhou Corporation

项目名称：叙品设计苏州公司
项目地点：江苏省昆山市
建筑面积：1000 m²
主要材料：蘑菇石、木条、空心砖、白色乳胶漆
主案设计：蒋国兴
空间摄影：吴辉（牧马山庄空间摄影机构）

Project name: Xupin Design Suzhou Corporation
Project location: Kunshan , Jiangsu
Building area: 1000 m²
Main materials: Mushroom stone, batten, hollow brick, white emulsion paint
Project designer: Jiang Guoxing
Photograph: Wu Hui (Graze horse villa photograph organization)

叙品诞生于一个美丽的冬天，在许多人看来，南方的冬天潮湿而灰暗；而在我们看来，因为寒冷，我们才更要装扮人们的生活空间；生活之所以美丽，是因为我们拥有美好的愿望、无穷的想象力，抱着这样的愿望，我们开始了我们的梦想……

空间设计中，在色彩和布局上，都跳脱传统，独树一帜，一味堆砌白色元素，巧妙地运用白色以及与其他颜色的配合，白得很有意境，营造出"此时无声胜有声"的氛围。

借用传统园林设计中"欲扬先抑"的手法，低调的门、狭窄的走廊、隐蔽的入口，却在转身的一刹那豁然开朗，别有洞天。鹅卵石夹道，古风荡漾却清新怡人，走过长廊，尘世的烦恼也慢慢抛之脑后。

Xupin was born in a beautiful winter. The winter in southern China is moist and dark in many people's view. But in our opinion, we should decorate people's living space because of coldness. The reason why life is beautiful is that we have good wishes, boundless imagination. We start our dreams with such a wish.

The color tune and layout of space is unique and different from tradition. White is the main tune, its application with other colors create such an atmosphere of silence.

The design uses the contrast skill of traditional garden design, such as concise door, narrow corridor, and covert entrance and so on. You always find something surprising at the corner. The ground of long corridor paved on cobbles is retro; you will forget all annoyance of life when you walk on it.

绿色象征着生命，白色象征着优雅，从硬装到家具陈设，配色都协调统一，体现出对细节的完美追求。人和物在这里共存，融为一体。

"梦幻"的主旨对中式文化进行了很好的现代性表现，空间总会在细致入微的地方给你意想不到的美妙感受，简洁纯净的主题始终贯穿整个案子。

办公室是人们从事脑力劳动的场所，员工的情绪、工作效率常常会受到来自环境的影响。而在叙品设计的这间新办公室中，轻松愉快的色彩、别致巧妙的创意，再加上空气中弥漫的茶香味，所有这些都可以让工作人员在放松的心情下完成工作，从而有利于工作效率的提高。

这不仅是一种健康的工作方式，更体现了人们积极的生活态度，在这个物欲横流的年代，希望本案的设计能让烦躁的人们重新寻找到最初的质朴，静下心来快乐地工作，同时能感受真正的生活。

Green is a symbol of life, white is a symbol of elegance, the color from hard outfit to furniture is in harmony, showing the pursuit to perfect details. People and objects, exist in harmony.

The theme of dream embodies the Chinese culture greatly in a modern way, space will surprise you at some subtle point, and the other theme of project is concise throughout the whole project.

Office is the public space for mental workers. The surrounding environment affects their emotion and working efficiency. In the office of Xupin Design, the pleasant color, unique creation and scent of tea make staffs relaxing, which it promotes the working efficiency in some degree.

This is not only a healthy way of working, but also reflects people's positive attitude towards life, in this materialistic age, the design of this project makes fretful people find the original simpleness, get down to work happily and feel the real life.

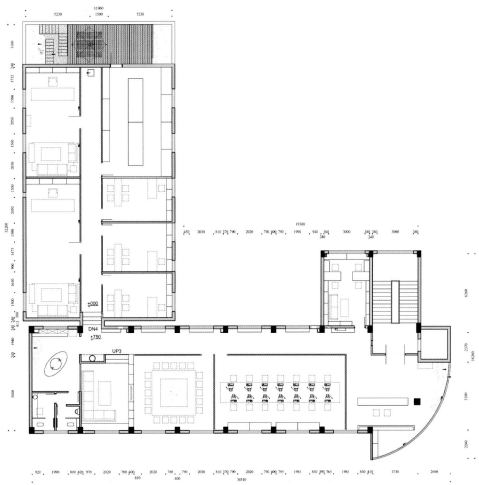

121

不诤素食馆
Buzheng Vegetarian Restaurant

项目名称：不诤素食馆
项目地点：新疆
建筑面积：750 m²
主要材料：黑色荔枝面大理石、木拼条、夯土、黑色高亮砖
主案设计：蒋国兴
空间摄影：吴辉（牧马山庄空间摄影机构）

Project name: Buzheng Vegetarian Restaurant
Project location: Xinjiang
Building area: 750 m²
Main materials: Bush-hammered black marble, wood batten, rammed earth, black highlight brick
Project designer: Jiang Guoxing
Photograph: Wu Hui (Graze horse villa photograph organization)

素食，表现出了回归自然、回归健康和保护地球环境的返璞归真的文化理念。吃素，除了能获取天然纯净的均衡营养外，还能额外地体验到摆脱了都市的喧闹和欲望的愉悦。悄然传播的素食文化，使得素食越来越成为一个全球时尚的标签。素食，已经成为一种全新的环保、健康的生活方式。

本案是一个主打素食的餐饮空间，分为两层，一楼为接待收银区，二楼是包间卡座区。

Vegetarian diet is a conception of returning nature and health, and protecting environment. You can absorb balanced nutrition and experience the pleasure of keeping away from city's noise and desire. Vegetarian diet culture is becoming more and more fashionable nowadays, and it spreads a kind of new healthy and environmental life-style.

The project is a vegetarian space that has two floors. The first floor is for reception counter, and the second is for private rooms.

进入大厅，一整面的白砂岩墙面，中间设计了一个小小的六角窗造型，两边摆放着简洁的中式椅子和落地灯，既简洁又古典。内凹的壁龛在灯光的照射下发出淡黄色的光，其他三面墙均以木格作为装饰，雅致又通风。竹编的弧形顶面看起来像中式走廊的屋檐。

往里面走，斧刀石的墙面，粗犷又大气。等待区的中间规划了一处水景，有山有水还有小船，顶面还飘着一片云彩，这样的空间使人想不静下来都难。透过六角窗，又可以若隐若现地看到前厅。黑色方管和铁板组合的层架插满了不规则的小木块，很好地起到了装饰的作用。

There is a white sandstone wall in lobby; at the center of it is a little six-angle window. Chinese-style chair and floor lamp are put at both sides which is concise and classic. The concave wall-lamp glows yellowish light. The other three walls are decorated in grillwork, which is ventilated and tasteful. The arc bamboo weaving ceiling is like roof of Chinese corridor.

Entering the inside, you can see the rough wall whose veneer is made of culture stone. There is a waterscape in the center of waiting area. The waterscape is a combination of fake mountains, water, boats and mimical clouds, and you can relax absolutely and be quiet there. The anteroom can be seen through the six-angle window indistinctly. Black square tubes and iron laminates are inserted in a lot of irregular and small blocks, which are beautiful and plain.

125

服务台延续了隔断的造型，木质的小花格静静地立在那里，黑色层板架上摆满了红酒，玻璃层板上面散发出淡淡的黄光。墙面是一幅巨大的黑白水墨画，顶面的设计延续了前厅顶面的造型。

楼梯下面做了一个枯山水景观，白色的粗沙，尖尖的石头，挺拔的枯树，与水景形成鲜明的对比，一动一静，一实一虚。

二楼走道采用了木拼条的造型，墙顶结合，地面采用了亮面的黑色地砖，内凹的壁龛在灯光的照射下透出微弱的光芒，土陶罐随意摆放着，整个空间没有多余的灯光，简洁又素雅。卡座区延续了一楼的隔断造型，顶面设计了窄窄的天窗，透过玻璃，微弱的月光洒进室内，偶尔还能看见点点繁星。

Service counter also applies partitions which is a kind of wooden grillwork. Some bottle of wine are put on the black board under yellow light. A giant Chinese landscape painting is on the wall, and the ceiling continues the style of anteroom.

There is dry landscape under stairs. White rough sand, sharp stones, straight dry trees are contrasted with flowing water.

The second floor corridor use wood battens. Ceiling matches with wall where concave lamps are with gentle light. Black tiles is paved on the ground, and some steans are placed casually. There is no more extra light in the space; the whole atmosphere is concise and elegant. Leisure area follows the style of wooden grillwork on the first floor. On the ceiling there is a narrow window, through which moonlight pour in through the window, and you can see stars when you look up.

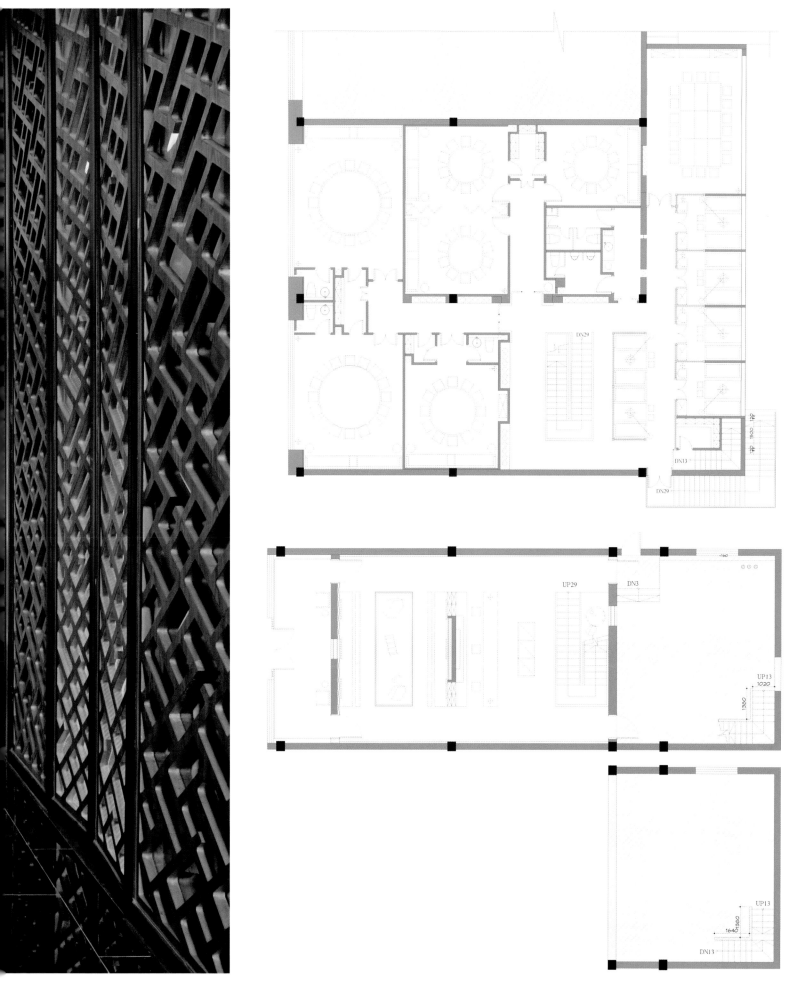

包间采用了条砖、斧刀石、泥色海藻泥、黑白壁画等质朴粗犷的材质，搭配简洁的中式家具，点缀着白桦木的装饰，营造出一种自然、素雅的空间氛围。

洗手区的墙面贴满了粗犷的斧刀石，地面是亮面的黑色地砖，形成鲜明的对比。台盆旁边的枯木在灯光的照射下愈发衬托出空间的宁静，卫生间则采用了黑色的荔枝面大理石，低调而深沉。

Decoration elements of private room are bricks, culture stones, brown seaweed mud, and black and white murals, natural and plain, matching with concise Chinese furniture and white birch.

Wall of washroom adorns with rough culture stones, which forms an intensive contrast with the ground paved with black bricks. Dry woods beside the basin make a sense of tranquility in the light. Bush-hammered black marbles are used in toilet, which is ordinary but mysterious.

135

花食间
Hua Shi Jian Restaurant

项目名称：花食间	Project name: Hua Shi Jian Restaurant
项目地点：新疆乌鲁木齐	Project location: Urumqi, Xinjiang
建筑面积：720 ㎡	Building area: 720 m²
主要材料：木拼条、银镜、大理石、绿植、实木地板	Main materials: Wood batten, mirror, marble, plants, wooden floor
主案设计：蒋国兴	Project designer: Jiang Guoxing
空间摄影：吴辉（牧马山庄空间摄影机构）	Photograph: Wu Hui (Graze horse villa photograph organization)

　　花食间位于新疆乌鲁木齐公园北街,覆盖有大量的植被树木,周围有亭台楼阁,使其仿佛置身于江南。餐厅选址闹中取静,为顾客提供了一个舒适的就餐环境。

Hua Shi Jian Restaurant locates at Park North Street, Urumqi, Xinjiang. The place is like the Chinese traditional south towns which are covered by plenty of plants and a few kiosks. Located in a quiet place, the restaurant provides a comfortable place for its customers.

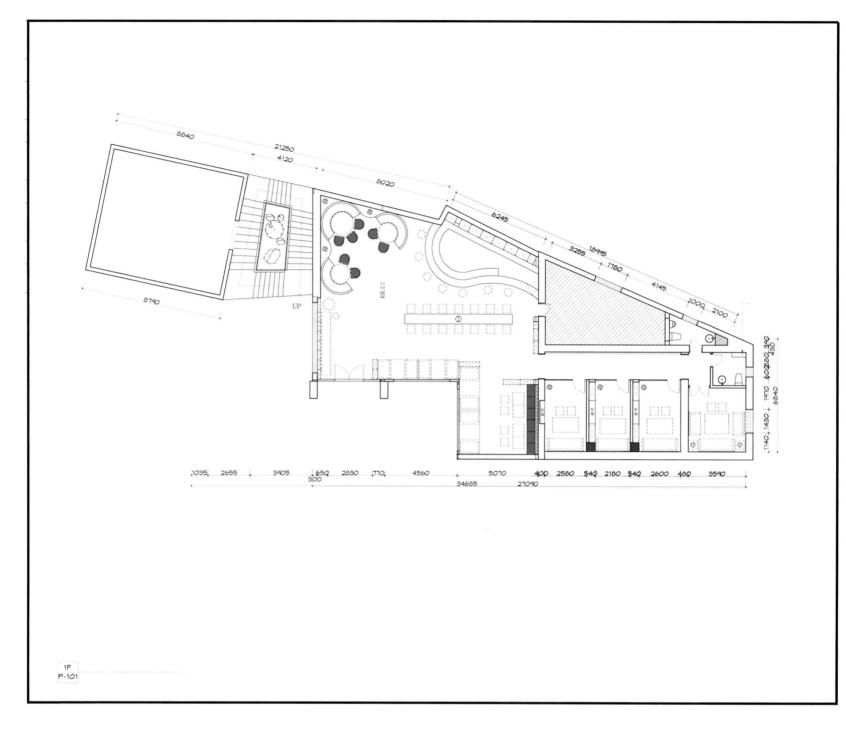

本案中艺术与智慧高度统一，设计师摒弃了以往的设计风格，打造了一个全然一新的世界。空间沉静、复古、富有文化气息而又时尚鲜活。

The project unifies art and wisdom to create a brand new restaurant, which features western-style food. The design style of the restaurant contains a lot of special elements, creating a quiet, cultural and fashionable space.

　　明末清初思想家王夫之曾言："凡虚空皆气也，聚则显，显则人谓之有；散则隐，隐则人谓之无"。"空也是无"间分和合便形成空间，便藏风聚气，可谓"无中生有"。在空间分割上，设计师没有刻意地强调某一部分，尽量做到开敞、通透、一目了然。

　　进入大厅，顶面为波浪形镜面马赛克，风起云涌，云卷云舒，映衬墙面起伏的沙丘似的"草原"，风云变幻，宽广无界。木纹大理石地面与绿植、天空构成气势恢宏的画卷，缓缓铺开。顶面一条铁丝网状的灯带，贯穿整个空间，蜿蜒曲折，如天边飘来的云彩。

　　Wang Fuzhi, who lived in the late Ming and early Qing Dynasty said that the void is a kind of energy; people can see it when it gathers together, and can't see it when it scatters. The void is nothing. Space is formed by the gathering and scattering of energy. It is a process of creating. The designer divides the space naturally, open and transparent.

　　When you come into the lobby, the waved mirror mosaics like the wind and cloud will come into your sight. A painting of waved dune-like boundless grassland on the wall is mysterious and exciting. Wood-grain-marble floor with plants and ceiling makes up a grand scene which is emotional. An iron net-shaped light strip goes through the space, which is like floating clouds.

服务台同样为弧形,与顶面、墙面相呼应,呈波浪形,宛如白色丝带在空中飞舞,柔软而光滑。

本案在软装方面,运用了大量彩色元素,使得整个空间动感十足。蓝色的模特灯、粉红的餐椅,无一不透露着时尚的俏皮感。

右面为包间,包间采用米灰色墙面和镜面马赛克顶面,让空间沉静如秋水,加上软装饰用色的搭配,增添了许多跳跃感。壁炉的火焰,让人瞬间就能感到温暖的气息。

Arc-shaped service counter has the same elements with ceiling and wall, which is like white soft and smooth silks floating in the space.

The ornament is colorful to vivify the whole space. Blue model light and pink dining chairs are fashionable and active.

There are private rooms on the right side. Khaki wall and mirror mosaics on the ceiling make a sense of peace and contrast with colorful ornaments. Flame in fireplace warms up the room.

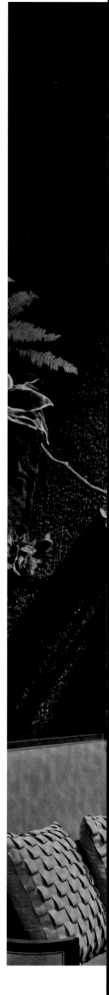

向里走进过道，绿色植物从左侧墙面的林木间长出，一呼一吸都是清新的空气。植物在这里尽情地舒展生长。

右面是散座区。木色做旧的墙面，衬以白色的画框，简洁大气。封闭的空间加了两个假窗，给人以无限遐想的空间，仿佛可以从那里透进阳光，左边墙上点缀以绿植，营造出一个温暖舒适的就餐氛围。

推开洗手间的门，让人有种置身于山水之间的错觉。

Moving along the corridor, green plants grows up on the left side wall, providing fresh air.

The retro wall surface and white frames in leisure area on the right side are concise. We design two fake windows in closed space, which stimulates imagination as if there is sunlight through the window. It makes a warm and comfortable atmosphere with plants on the left wall.

Opening the door of washroom, there will be a feeling in beautiful hills and waters, fantastic and quiet.

般若世界·孔雀餐厅
The World Peacock Restaurant

项目名称：般若世界·孔雀餐厅
项目地点：新疆乌鲁木齐
建筑面积：920 ㎡
主要材料：白色木纹石、白色麻织壁纸、浅色木拼条
主案设计：蒋国兴
空间摄影：蒋国兴

Project name: The World Peacock Restaurant
Project location: Urumqi, Xinjiang
Building area: 920 m²
Main materials: White wood grain stone, white linen wallpaper and light wood batten
Project designer: Jiang Guoxing
Photograph: Jiang Guoxing

"珠花冕，翠耳环，锦丝织就绿罗衫，
玉佩金钏镶彩扇，灵眸秀姿气宇端。
影婀娜，舞翩跹，程派低回声婉转，
慢摇莲步轻顾盼，清雅富贵好悠闲。"

"Bead crown, emerald earrings, silk brocade on the Green Shirt,

Jade and gold decorations inlaid colored fan, bright eyes and good figure shows aristocratic manner.

Shadow graceful, dance tripping, sweet voice feature Cheng style,

Walking slowly and gently, with a charming look, elegant and leisurely."

古往今来孔雀都被视为吉祥、善良、美丽、华贵的象征。如果天鹅代表了西方,那么美丽的孔雀则能代表东方。

本案设计师以孔雀为主题,意境作景,打造一个明朗、大方、纯净、高贵的餐饮空间。项目位于乌鲁木齐市人民公园西侧,紧挨着这个闹市中安静的公园,因此以静为出发点,体现中式意境。整个空间以白色为主色,以孔雀元素的孔雀蓝为辅色,孔雀蓝与白色搭配,干净而安静;用纯净的手法、普通的材料营造动人的气质,打造出安静而祥和的就餐空间。

Since ancient time, peacock is a symbol of good luck, kindness, beauty and luxury. If the swan represents the West, then the beautiful peacocks represent the East.

The designer uses peacock as the theme in this case, and in artistic scene, to create a clear, natural, pure, and noble dining space. The restaurant is located on the west side of Urumqi People's Park, a quiet park in the downtown area, and thus starts with quietness, reflecting Chinese artistic conception. White is the main color throughout the space, with peacock blue color, which is clean and quiet. The project uses simple technique, ordinary materials to create beautiful temperament, deriving a quiet and peaceful dining space.

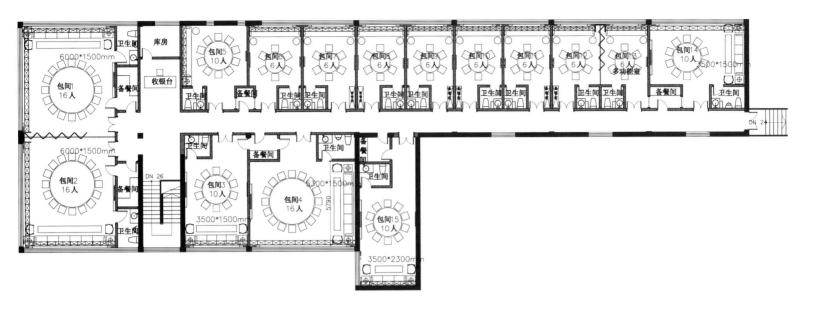

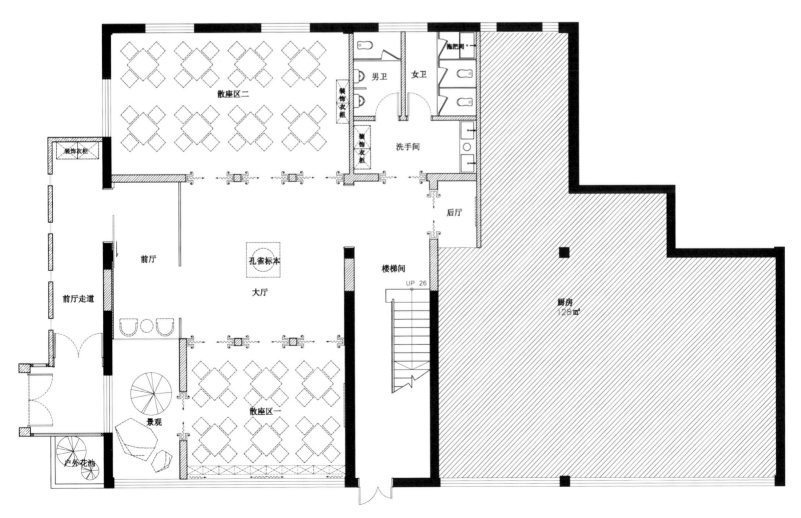

从餐厅入口步入大厅，映入眼帘的是层次错落的白色木座格栅造型，格栅的空隙间隔使人产生想一探究竟的冲劲，显现出神秘的空间感。进入餐厅大厅，即可看到地面白色的木纹石材与墙面白色的麻织壁纸和白色的吊顶相得益彰。在色彩灯光的大厅中央悠然矗立着一只婀娜多姿、翩翩起舞、身着蓝绿色罗衫的孔雀标本，恍惚间这安静而祥和的就餐空间已然那样欢呼雀跃。大厅两侧是散座就餐区，进入就餐区的门洞上的白色木作线条带有一圈灯带，金黄色的光具有醒目的指示作用。门洞中间夹着麻布色纱帘，将就餐区与大厅明显地区分开来。就餐区仍采用白色的麻织壁纸，白色木质椅子与孔雀蓝坐垫、靠背相搭配。孔雀蓝既有古典的高贵、典雅韵味，又有流行的明艳、纯净特质。这个颜色是遥不可攀的神界色彩，是除了金银以外的一个特殊色。因此赋予了整个明朗的空间一种特有的气质与热情。再加上墙面的孔雀元素的壁画，使整个空间更加生动起来。包厢部分集中在餐厅的二楼，大包厢的墙面、吊顶延续了就餐区的白色明朗的元素，简单的吊顶与简洁的墙面线条，使圆形餐桌与奇特又新颖的羽毛灯更引人注目，让这独具魅力的包厢显得更加清新、灵动！

From restaurant entrance into the hall, you can see a white well-arranged wooden grid the gap between which make people want to explore the mystery of space. At the restaurant hall, the ground is paved with white stones, and wall surface is covered with linen wallpaper. The ceiling is white. At the center of the hall, there is a beautiful peacock specimen, dancing gracefully and painted with bright blue and green. The hall is flanked by dining area, and doorway to it lines with a circle of lights, The golden light has indicative function. Linen curtains are set between dining area, showing clear boundaries with the hall. Dining area use white linen wallpaper, white wood chairs and peacock blue cushion, creating a feeling of classical and noble, elegant and popular. This color is a special color in addition to gold and silver, providing the space with specific temperament and passion. Coupled with the murals with peacock elements, the space is more interesting. Box part is on the restaurant's second floor. The wall and ceiling continue to use the white bright element of the dining area. Simple suspended ceiling, simple wall lines, remarkably circular dining table and strange and novel feather light make this charming box fresh and cheerful!

本餐厅以东方孔雀为主要元素,并引用现代中式元素,让人远离繁忙、喧嚣的世俗,在这充满文化气息、舒适安逸的环境里找到属于自己心中的一片宁静。

This restaurant takes oriental peacock as main elements, adorning with modern Chinese elements, to make people move away from busy and noisy world, finding their own comfortable place to stay calm.

竹溪一号长春路店
No.1 Zhuxi, Changchun Road

项目名称：竹溪一号长春路店
项目地点：新疆乌鲁木齐
建筑面积：1000 m²
主要材料：做旧红砖、素色水泥墙等
主案设计：蒋国兴
空间摄影：吴辉（牧马山庄空间摄影机构）

Project name: No.1 Zhuxi, Changchun Road
Project location: Urumqi, Xinjiang
Building area: 1000 m²
Main materials: Old red brick, plain concrete walls
Project designer: Jiang Guoxing
Photograph: Wu Hui (Graze horse villa photograph organization)

本案定位为工业与田园的混搭风格,没有刻意地追求复杂的造型,而是以最简单的元素和最质朴的材料打造出无限丰富的效果,使空间更富有张力。它要简单,有着工业时代的横平竖直;它要田园,却绝对和"花花草草"没有关系。复古的工业风、最简单的线条、最质朴的材质,营造出别具一格的餐饮氛围。

The case is positioned as a mix of industrial and rural style. Without deliberate pursuit of complex shapes, simple elements and materials create a rich effect to make the space have more tension. It should be simple, with industrial-age's horizontal and vertical elements; it should be pastoral, but has no common with "flowers and grasses". Vintage industrial style, the simple lines, the simplest materials and unique style create a special dining atmosphere.

在平面布置上，规划了大厅和包厢，通过声、形、光将散座区有序地分散在大厅中，动线流畅。在色彩运用上，以灰色、黑色、土黄色为主，搭配浅色的家具，营造了一种别具一格的用餐环境。

In terms of layout, the designer plans the hall and VIP rooms, and sound, shape, light makes dining area well arranged throughout the hall with smooth line. On the use of color, the designer uses grey, black, khaki, with light coloured furniture, to create a unique dining enviroment.

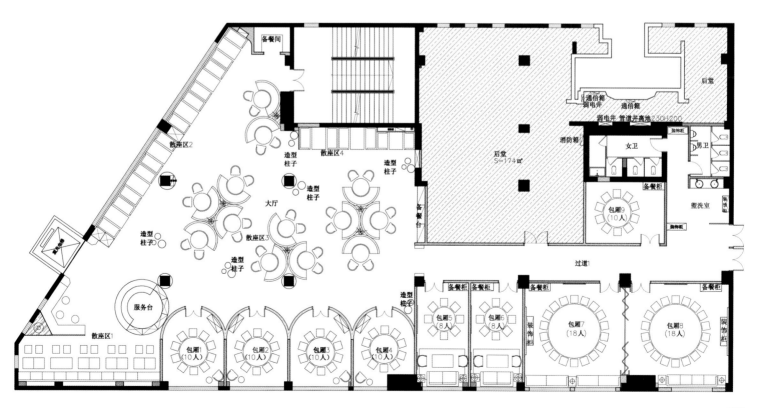

进入大厅，首先映入眼帘的是圆形的服务台，与顶面圆形的吊架相互呼应着。服务台外立面贴着粗犷的蘑菇石，外置一圈黑色的铁丝网造型，简单、质朴，在射灯的照射下，别具工业时代的风格。服务台的旁边是等待区，在用餐高峰期，为客人提供了一处很好的休息场所，人性化的设计提升了餐厅的服务品质。休息区的后面是一排散座区，墙面装饰了各种各样的连环画，活跃了空间的氛围。服务台后面是4个连包，采用了黑色的铁丝网镂空隔断，延续了服务台的设计元素，使空间隔而不断。大厅的中央是圆形的散座区，顶面吊着圆形的铁丝网吊架，与地面相互呼应。土陶罐饰品装饰着大厅的每一个角落，铜质的马赛克点缀着每一根柱子。

Entering the hall, the first thing to be seen is a circular service desk, echoing with top round hanger. The desk has a bold mushroom stone facade, circled by a black wire-netting, simple and rustic under spotlights, showing the unique microcosm of the industrial age. Next to the desk is the waiting area, which offers guests a good resting place at the peak time of dining, and user-friendly design enhances the quality of the restaurant service. The rest area is behind a row of seats, where walls are decorated with a wide variety of comics, activating the atmosphere of the space. Behind the desk are 4 linked rooms, where a black wire-netting hollow partition is used to continue the design element of the service desk and to ensure the continuity of the space. The center of the hall is a round sitting area, hanging a round wire hanger. Earthenware jars can be seen at every corner of the hall, and copper mosaic tiles decorate pillars in the room.

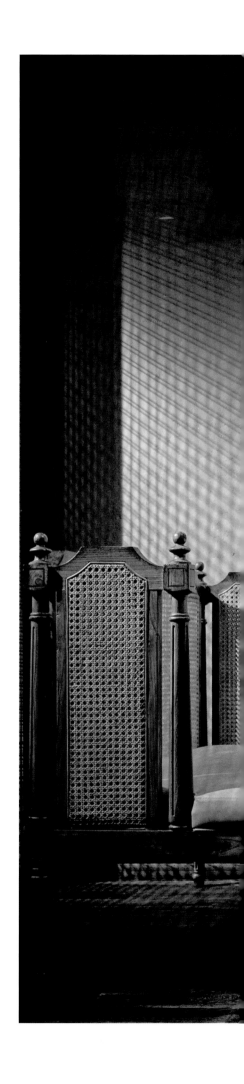

　　包间没有复杂的造型，复古条砖搭配铁艺的灯具，木本色的家具，散发出一种质朴、自然的韵味。

　　通往走道的垭口采用的是铁锈板装饰的大门套。走道的一面墙采用素色的水泥打造，质朴、粗犷。另一面墙则是木框与灰镜组合，在视觉上延伸了空间。

　　"碧云天，黄叶地，秋色连波，波上寒烟翠。"乌鲁木齐的秋天总是那么美，悠然的午后，在这里享受下午茶时光，脑海又响起了回忆的旋律……

There are no complex shapes in the private rooms. By using retro bricks, iron lamps and wood furniture, the rooms have a simple and natural feeling.

Pass leading to the walkway uses door pocket decorated with rusty plates. Cement modeling wall is plain, simple and bold. Wall on the other side is a combination of wooden frames with grey mirror, visually extending the sense of space.

"White clouds are all over the sky. Yellow leaves are everywhere. The autumn scenery is reflected into the waves of the river, which is shrouded in chilly green." The autumn of Urumqi is always beautiful. In leisurely afternoons, you can enjoy tea time and remember all the good times.

三仟健康
Sanqian Health

项目名称：三仟健康
项目地点：新疆乌鲁木齐
建筑面积：1008 ㎡
主要材料：原石、雅士白、黑板岩、实木地板、做旧红橡木、锈铁板
主案设计：蒋国兴
空间摄影：吴辉（牧马山庄空间摄影机构）

Project name: Sanqian Health
Project location: Urumqi, Xinjiang
Building area: 1008 m²
Main materials: Stone, Ariston marbles, black slates, wood floor, red oak, rusty plates
Project designer: Jiang Guoxing
Photograph: Wu Hui (Graze horse villa photograph organization)

多年前,在电视上看过一个头戴花环的女子,静坐在瓦蓝色的大海边练瑜伽。她气定神闲、静如处子,舒缓、绵长、柔软、曼妙的肢体语言,透着一种莫可名状的神秘之美,无形之中把人带入一种妙不可言的秘境中。从此,心中对瑜伽有一种梦幻般的神往。

瑜伽是梵文Yoga的音译,意思是"相应"。与什么相应呢?与境相应,与行相应,与理相应,与果相应,与机相应。从通常的意义来说,瑜伽有"抑制心的作用",从这个意义上说,瑜伽与禅是同义语,也可以说禅是瑜伽的继承与发展。

Many years ago, I saw a woman on TV wearing a Garland, sitting by blue sea and practicing Yoga. She looks quiet, calm and deliberate, using a slow, soft and graceful body language to reveal an uncanny sense of mysterious beauty of invisible people in a fantastic fam. From then on, Yoga absorbed me.

Yoga is a transliteration of the Sanskrit word meaning "corresponding". And what does it correspond with? With the contexts, the lines, the principles, the results, and the opportunities. From the usual sense, Yoga is "control your heart". In this sense, Yoga and Zen are synonyms, in other words, Zen is the inheritance and development of Yoga.

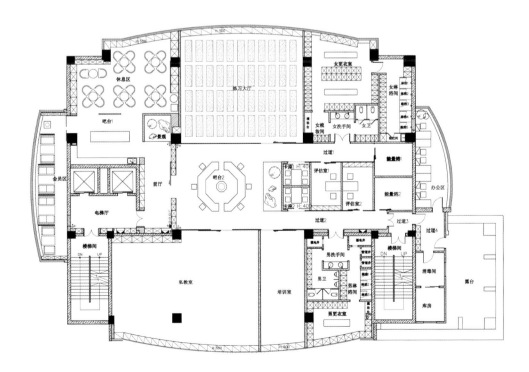

瑜伽是东方心灵治学，通过调试内心活动，可清除人潜意识里的杂念，消除烦恼，是一种减压和心灵美容的良方。

瑜伽馆当然应该是恬淡、宁静、自然、清新的。带着自然的禅意，但这些抽象的概念要体现在空间中，并让观众都接收到，却不是一件容易的事。所以，伪设计、伪禅意的空间特别多。三仟瑜伽馆的设计可谓轻松、自然、游刃有余。

从电梯间出来，墙面使用了土黄色火山岩，材质简单地平铺，拒绝繁复，回归简洁。

进了门，是木拼条的墙面。木材无论从视觉上还是触觉上都给人一种温暖的感觉，墙上用锈铁板做的LOGO，背后透光。地上随意摆放几盏蜡烛灯，透着一种天然拙朴的气质，柔和刚原来可以结合得这么浑然天成。门口有一排木质的换鞋凳，凳子下设计了放鞋子的小空间，用麻灰色的布遮掩着，背后打着暖暖的光，瞬间有一种温暖的归属感。

主入口的小景观，地上是一块大大的石头，未经雕琢，石头背后是一棵白色的枯木，顶上飘着轻轻的一片云，带着简单又纯净的自然气息。走进去是原石做成的吧台，石头棱角分明，和周围的木色相互衬托，这种宁静让从喧嚣世界走进来的我们瞬间安静了下来。透过旁边的木格可以看到休息区，几个小圆桌，原木色的椅子，等待的客人坐在那安静地看看书，低声闲谈。背后的墙面设计得很特别，细细的木格子围成的框，里面塞满了大小不一的木块，粗糙的截面未经打磨，散发着原木的清香。

吧台区和接待台一样用了原石和锈铁板，八角形的设计寓意八面玲珑，也形成了一个围合空间，可以最大限度地照顾到各个方向的顾客。"八"在中国文化中本就占据很高的地位，古代八卦图也是正八边形，寓意阴阳调和、万物的变化轮回。天花上也是很巧妙地用了白色的镂空铁片编织，疏密有致，不至于沉闷，也不至空旷，取舍有度。两边是木质隔断，布纹玻璃背后透着浅浅的灯光，让后面的空间给人一种神秘感，很想轻轻推开门看个究竟。

Yoga is also a kind of psychology in the oriental country. People adjust mood and remove all the worries through it. it's a good method for relieving pressure to create a healthy emotional environment.

On mentioning the place "Yoga club", such words like peaceful, natural, quiet, and fresh immediately come into mind. People realize Zen in nature, but it is challenging to apply these abstract elements of Zen sense to space design and make audience feel it at the same time. And therefore, there are many unsuccessful space designs that want to show Zen sense. The space design of the "Sanqian" Yoga club is natural, relax and skilled. We can feel that the designer is very sure of his ability, and the idea about design has considered for a long time.

When you come out of elevator, you can see the concise decoration of yellow volcanic stone of the wall.

When you come into the door, the use of batten of the wall give you a visual feel and touchable feel of warm for the space. There is a logo made of rusted board on the wall, light penetrates behind it. Some candle light set casually on the floor, making a sense of natural and rustic. Soft and hard are matched so well. A row of wooden bench for changing shoes set beside the door. Under the stool, there is a small space for putting shoes covered with linen and a little warm light on the back. The whole atmosphere is familiar.

A small plant decoration is set at the main entrance. There is a big natural stone on the floor, a white dry wood is set behind it and a cloud decoration floating over it. The whole space is simple, clean and natural. You can see a bar counter which is made of original stone when you go inside. The stone is clean-cut, and is matched with the color of the wood. Peace of space make people leave away the blundering world and find inner peace. You can see the lounge through wood frames beside. Some small round desks, wood chairs are prepared for waiting guests who can sit down and read books or chat. The wall on the back is designed specially. Some slim lattices form a frame, and all kinds of wooden blocks are filled in lattices. The cross-section of blocks is rough and the smell of log is sweet.

Bar counter and reception counter both apply materials of original stone and rusted board. Eight-square space design means everything goes well and make enclosed area. Customers in all directions are serviced well in this area. The number of eight occupies an important position in Chinese culture. Bagua image of ancient China means harmony between yin and yang, everything's change and return. White and hollow-carved iron nets decorate on the ceiling ingeniously. The density of hollow-carved is changed irregularly, and therefore, the ceiling looks active, abundant and rhythmic. There are wooden partitions on either hand. Soft light penetrate through wire glass, it makes the back space mysterious and make people want to get in to see truly what there is.

高温瑜伽室、私教室的顶面延续了过厅的白色编织铁片,一面为通铺的镜面,对面是木拼条,侧面则是白色木格栅的移门,整个空间素雅得犹如未经世事的仙子。灯光透过白色铁片,星星点点撒在这个安静的空间。赤脚走在温暖的木地板上,在蒲团上坐下,闭上眼睛,思绪仿佛暂时停止,什么也不想,只有柔和的风吹过窗户,吹起麻灰色的窗帘,吹到脸上。

卡座及评估室做成和室的感觉,藤编壁纸与木格栅搭配,并在侧面设置了一面镜子。

The ceiling of Yoga chamber with high temperature, and the ceiling of private classroom both extend the decoration of white and hollow-carved iron nets. On the one side there is a whole mirror wall. And on the other side, wooden battens decorate on the wall. A horizontal sliding door decorated on white wooden batten is set on the side. The whole space is pure and elegant like an angle. Light penetrate through the white iron net and fill this quiet space. When you step on warm wooden floor with bare feet, sit in the futon, close your eyes, and concentrate on around environment. You will feel soft breeze blow in the room, through window and linen screen, and cross the face, making you forget everything.

Booths and evaluation room are designed in Japanese-style. Rattan wallpapers are matched with wooden battens, a bright mirror sets on the side.

洗手间墙面运用白色石材，地面为灰砖。洗手台下方凹空，可用来摆放装饰品，这种设计同时提升了空间品质。

洗手间背后设置了淋浴间和更衣室，为顾客提供了一个舒适的环境。

Bathroom walls use white stone and the ground use grey brick, and concave below the sink to put ornaments, enhancing the quality of the space.

The shower and locker rooms are set behind the bathroom, providing a comfortable environment for customers.

沙味传奇
Sha Wei Legend

项目名称：沙味传奇
项目地点：新疆乌鲁木齐
建筑面积：720 ㎡
主要材料：铁片编织、水泥砖、花砖、瓦片、黑色铁丝网等
主案设计：蒋国兴
空间摄影：吴辉（牧马山庄空间摄影机构）

Project name: Sha Wei Legend
Project location: Urumqi, Xinjiang
Building area: 720 m²
Main materials: Iron weaving, concrete bricks, tiling, tiles, black wire
Project designer: Jiang Guoxing
Photograph: Wu Hui (Graze horse villa photograph organization)

本案位于新疆乌鲁木齐世纪金花时代广场，一座新型的城市综合体，以绝对的地标建筑身份，牢牢占据着乌鲁木齐繁华的核心商务区，区域内吸引了近4000家大中型企事业单位云集，成为人流、物流、信息流高度集中的发达商务区。

目标客群分析：周边各大企业单位的工作人员；周边学生、老师；人来人往的客流人群。

此项目为改造工程，设计师必须要在尽可能保留之前的设计基础上进行加工创造，还需控制成本，这在工程总价与设计要求上都对设计师的各项素质提出挑战，必须用最低的造价打造最完美的效果。

This case is located in Century Ginwo Plaza in Urumqi, Xinjiang. It is a new urban complex, and an absolute landmark status dominating the most distinguished downtown central business district of Urumqi. It has attracted almost more than 4,000 large and medium enterprises, and becomes a developed business district with high-concentrated people flow, goods flow and information flow.

Analysis of target group: staff of surrounding companies; students and teachers; passengers.

The project is a renovation project. The designer must work on the project with limited cost on the basis of previous design, which challenges the designer from the aspect of lessening project cost and meeting the design requirement.

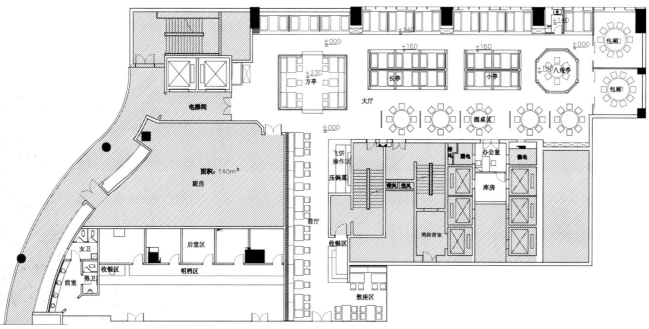

餐厅外围采用红色竹编材质,柔软的质地从墙面延伸至室内顶面,显得火辣妖娆,异域风情十足,不禁使人联想到西域美女曼妙的舞姿和西域人热情爽朗的性格,也贴合了餐厅的主题——新疆美食。

The exterior of the restaurant uses bamboo material painted in red, from the ground to the ceiling, creating a exotic style and associating with dances of beauties and hospitality of people in the west part of China. The design also fits the themes of the restaurant – Xinjiang cuisine.

左边为明档，地面同样采用了红色系拼花砖，一直延伸至明档服务台，仿佛从地面生长出来一样。与之呼应的是从外墙生长进来的红色竹编造型，这种曲面的造型柔化了空间。设计手法上采用无边框的设计，旨在营造出广阔无际的新疆地貌特征和新疆人豪爽大气的性格特点，真是妙不可言。墙面使用灰色水泥砖，简单地平铺，质朴而又粗犷。细节之处，无一不透漏出设计师的用心。

散座区用黑色方管结合铁丝网分割。跳跃的用色简洁明快、富有跃动的青春气息，营造出一个轻快时尚、快乐活泼的餐饮氛围。

On the left of the restaurant is the open cooking area, and the ground also uses red mosaic tiles, extending to the service desk, as if it were growing from the ground. The red bamboo-weaving exterior wall extends into the open cooking area. The curved ceiling softens the space. The designer used frameless design to show Xinjiang's geographic features and the openness and generosity of Xinjiang people. The wall is paved with gray cement bricks, simple and rough. All of the above details reflect the desiegner's intentions.

Private rooms are divided by black tubes and wires. By using bright colors, fashionable and joyful dining atmosphere is created.

右边为就餐区，沙发使用了大面的积柠檬黄，犹如西域炙热可触的阳光，并配以彩色餐椅，犹如一条条舞动的艾德莱斯绸。值得一提的要属墙面的瓦片装饰了，土黄的颜色让人想到驼铃叮咚行走在无边的丝绸之路古道上，或是儿时下雨天顺着屋檐淅淅沥沥的雨滴。顶面保留了之前的廊檐造型，通过瓦片的衔接，使之显得厚重而又不失时尚。过道墙面采用同色系的仿木地板砖及驼色稻草漆，使墙面相互独立又相互统一。用最简单的材质，传达出设计最内心的声音，洗尽铅华，回归本真。你还可以一边品尝美食，一边欣赏餐厅专门准备的飞饼表演。

On the right of the restaurant is the dining area, with lemon-yellow sofas and colorful chairs, which look like dancing Ide Rice silk. The wall is decorated with tile ornaments, reminding you of the Silk Road with tinkles of camel bells or raindrop streamed down from the roof. The ceiling adopts the same design as corridor, and through the connection of tiles the whole space looks stately and modern. On the corridor, the wall uses wood-like floor tiles and beige lacquer. By using the simple materials, the design is well presented. You can enjoy delicious food and fly cake show at the same time.

餐厅保留了原始的地台及亭架,为它换了"新装",统一在整个环境里,朴实简约,未经雕琢而又意味深远。

卫生间的设计不同于公共空间,用黑白砖进行区分,时尚感十足。白色木拼条隔断,营造亲切舒适的氛围。

Dining room retains the original platform and booth, and decorates with a new style, having harmonious effect.

Bathroom design is different from public spaces, distinguished by black and white bricks, full of fashionable sense. White wood batten is used on the partition, creating warm and comfortable atmosphere.

千禧丽人整形医院接待中心
The Reception Center of Millennium Beauty Cosmetic Surgery Hospital

项目名称：千禧丽人整形医院接待中心
项目地点：新疆乌鲁木齐
建筑面积：324 ㎡
主要材料：火山岩、浅色实木复合地板、仿古爵士白大理石砖
主案设计：蒋国兴
空间摄影：吴辉（牧马山庄空间摄影机构）

Project name: The Reception Center of Millennium Beauty Cosmetic Surgery Hospital
Project location: Urumqi, Xinjiang
Building area: 324 ㎡
Main materials: Volcanics, light-colored wood floor, white marbles
Project designer: Jiang Guoxing
Photograph: Wu Hui (Graze horse villa photograph organization)

　　千禧丽人位于乌鲁木齐核心商业区时代广场，是一家专业的美容整形机构。
　　随着时代的发展进步，都市人的生活节奏逐步加快，在工作之余人们渴求一个能放松身心、放慢节奏、享受美的地方。这里将满足你的要求。

Millennium Beauty is located in the central business district of Urumqi. It is a company specializing in cosmetic and plastic surgery.

With the development of the times, the pace of people life is gradually accelerating. They want to find a place to relax and enjoy the beauty of life. That is the starting point of the design.

进入大厅,竹节似的墙面苍劲有力,竹子四季常青,寓意青春永驻。如千禧丽人的初衷。悠长的小径通向远方,瞬间使人忘记了城市的喧嚣,静静地享受这里的美好。顶部为麻布挽成的绳结灯,如少女的舞袖般轻柔飘逸。地面采用黑色花岗岩,反射"千禧丽人"的字样,别致典雅。

In the hall, the design of the wall uses bamboo concept, which means eternal youth, creating a vigorous and forceful effect. There is a long and quiet corridor, making people forget the hustle and bustle of the city. The ceiling lights are covered with voile and tied by linen knots at the bottom, looking like the sleeves of dress of girls. The ground is paved with black granite, reflecting the logo of "Millennium Beauty".

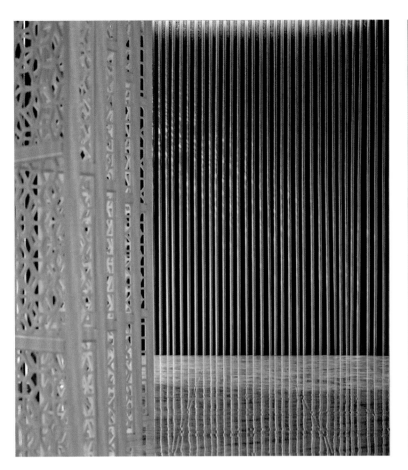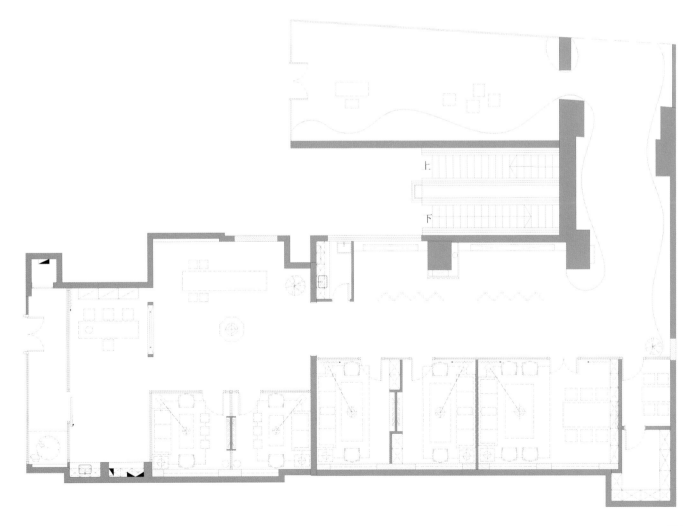

穿过八角形窗，是一个大休闲区，整体风格延续前厅。正前方的八角形柜与八边形窗相呼应，增添了一分秀丽。

踱步前行，左边为整形美容展示区，设计了镂空的折叠隔断，营造出一种朦胧美，若隐若现。另一道风景转角而遇，曲折蜿蜒的竹林使人身心放松。竹林尽头古筝声起，是一曲"高山流水"，此地虽无崇山峻岭，却有茂林修竹，确实为放松身心的好去处。

VIP室以江南风景为背景，描绘了一幅江南晚灯图。看到这幅景象眼前仿佛出现一位纤瘦的少女在江南背景下提一盏红灯笼走过，带动了微风，点亮了风景，给人以美的享受。

Through the octagonal window, there is a large recreation area and the overall atmosphere is continuation of the front hall. Octagonal cabinet and octagonal window make space more beautiful.

Then on the left part is the display area, where put folding partitions, creating a hazy beauty. Turning the corner, you will come across a beautiful scene – winding bamboo forest. At the end of it, sound of Guzheng flows out. That is "High mountain and running water", making people feel relaxed.

The VIP rooms take Jiangnan scenery as background, so beautiful as if you saw a slim lady carrying a latern and lightening up the scenery.

走进接待区，扑面而来的是一种质朴素净之感。服务台与地面统一采用白色大理石，简洁大气。背景用火山岩拼贴，用最简洁的材质来表达设计的主题——本真。服务台前方为发光的八角形窗，"八"意为八面玲珑。

隔断延续大厅的形式，取意竹节，节节高升，竹节必露，竹节镂空寓意虚怀若谷的心胸。隔断旁为休闲吧台，原木色的桌凳，可让人在此闲坐放松。吧台的后面设置了一个白色的木格栅柜子。

The reception area adopts a simple design. Reception desk and the ground use white marbles, concise and lofty. Volcanic rocks are paved on the background wall and are used to display the theme of the design – return to nature. At the front of the service desk stands an octagonal window "Eight" means being smooth and slick in establishing social relations.

The partition continues the design of the hall, by using bamboo elements. The moral of bamboo is to be successful with work, and the hollow part of bamboo means modest heart. Beside the partition, there is a bar, with wood-colored table and chairs. People can enjoy leisure time here. A white wood cabinet is set behind the bar.

合一茶道
Unity Tea

项目名称：合一茶道
项目地点：新疆乌鲁木齐
建筑面积：800 ㎡
主要材料：黑金沙大理石、水泥板、水泥砖、白色壁布、稻草漆等
主案设计：蒋国兴
空间摄影：吴辉（牧马山庄空间摄影机构）

Project name: Unity Tea
Project location: Urumqi, Xinjiang
Building area: 800 ㎡
Main materials: Black Galaxy marble, cement boards, cement bricks, white wall cloth, straw lacquer
Project designer: Jiang Guoxing
Photograph: Wu Hui (Graze horse villa photograph organization)

禅是东方古老文化理论的精髓之一，茶亦是中国传统文化的组成部分，品茶悟禅自古有之。设计师以禅的风韵来诠释室内设计，不求华丽，旨在体现人与自然的沟通，为现代人营造一片灵魂的栖息之地。整个空间以素色为主调，采用质朴的水泥板与水泥砖，为整个空间营造了一种大气磅礴的气势，以一种独特的姿态诠释着中式之美。

本案设计师将现代气息糅合东方禅意，演绎了一个优雅的品茗空间。在平面布置上分为两层，一楼规划了门厅、景观区，二楼规划了接待区、包间、厨房、餐厅、卫生间等。在色彩运用上，以黑色、白色、木本色为主色调，灰色为辅色调，再搭配一些花饰、器皿，让整个空间与茶道精神合而为一的同时又展现空间的全部功能和意境。

Zen is one of the theoretical essence of ancient Oriental culture, and tea is also an integral part of traditional Chinese culture. The tradition of drinking tea and understanding the essence of Zen is originated from the ancient times. The designer borrows Zen charm to explain interior design, aiming at embodying the communication between man and nature, and creating a resting place for people. Entire space, with plain color tone, uses simple cement boards and cement bricks, a kind of grand feeling pouring from it.

combining modern flavor with oriental Zen, the designer creates an elegant tea room. The space is divided into two floors. The hall and landscape area set on the first floor. While reception area, VIP rooms, kitchen, dining area and washroom are set on the second floor. On the application of color, the designer takes black, white and wood-color as main colors, supplemented with brown. Flower decorations and utensils are used to combine the space and tea spirit, displaying functions and artistic conception of the space.

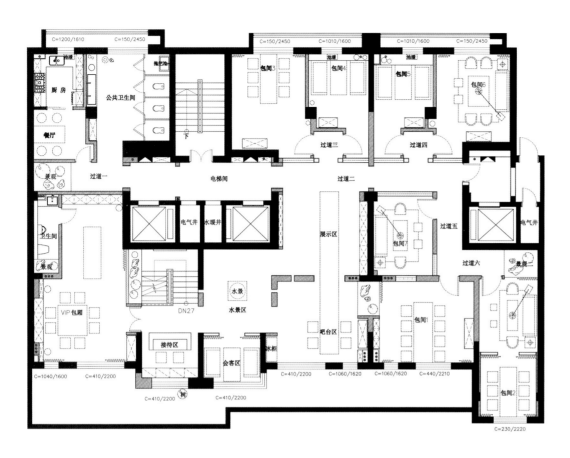
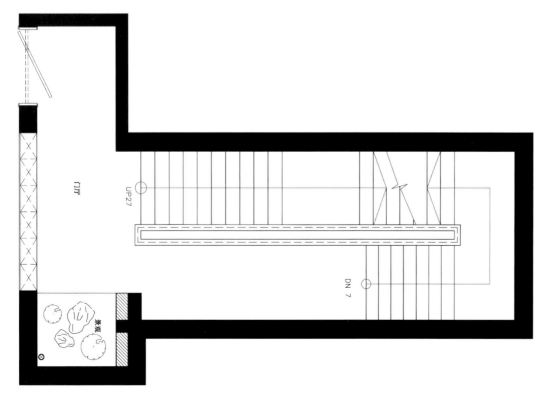

一楼的门厅规划了一个半通透的装饰柜，由黑色哑光方管和实木层板组合而成，后置钢化玻璃，隐隐约约，使室内空间与室外空间有一个视觉上的交流。素色的水泥板墙面、素色的水泥板栏杆、黑色的花岗岩踏步在灯光的照射下愈发使空间显得宁静、质朴。墙面的白色投影搭配一些小树枝装饰，还有旁边的白色枯木、石头装饰，禅的意境便油然而生，浑然天成。

二楼的接待台没有复杂的造型，黑色石材的台面，黑色的木作线条装饰着正立面，在灯光的照射下，透过磨砂玻璃，散发出淡淡的黄光。在会客区，设计师运用中国古典元素，特意设计了一个八角形的门框，意味着人生的八面玲珑。吧台区和展示区延续了这种元素，一张长长的桌子穿过八角门洞立在那里，桌面上摆放着大小不一的装饰陶罐，桌子下面随意地放着几只蜡烛，为空间增添了气氛。桌面上悬挂着一排黑色线条组成的面灯，既实用又满足了装饰的需求。红色的中式高柜给素色的空间增添了一些活力。枯山水的文化在这里得到了很好的应用，白色枯木、石头在灯光的照射下愈发显得精致。

At the hall on the first floor plan there is a semi-transparent decorative cabinet, made of black matte square tubes and wood board, at the back of which tempered glass is used; enhancing visual communication between the indoor space. Plain cement walls, plain cement railing and black granite steps make the space quiet and simple in the light. White projection on the wall, with some small tree decorations, dead wood and white stones, create an artistic conception of Zen.

Reception desk on the second floor do not use complex shape, instead, it has a black stone countertop, and black wood lines decorated the facade. Through the frosted glass, the desk emits a faint yellow light. In the sitting area, the designer employs classical Chinese elements, and specially designs an octagonal frame. Bar area and exhibition area keep using elements. A long table extends to the octagonal door, and decorative pots of different sizes are placed on the desktop and a few candles under the table, adding romantic atmosphere to the space. On the top pf the desk hangs a row of lamps with black lines, practical and decorative. Red chinese style cabinet adds some vitality to the space. Dry landscape is well used here. White dead wood and stones become more refined.

　　水景区抛弃一切矫饰，力求做到平淡致远，保留事物最基本的元素。素色的水泥板墙面、白色的投影、大小不一的陶罐、水景，形成一幅天然的具有诗情画意般的画面。

　　走道采用了质朴的水泥板和木拼条，追求表面的质感和肌理。为了弱化硬朗的材质，设计师还做了一些细节处理：走道尽头随意设置的枯木、石头景观，展示柜内的陶罐与茶具，墙面上的黑白挂画，这些细微之处的累积都让空间显得更为饱满。

　　包间没有复杂的造型，以素色为主，并使用米色的稻草漆、白色的壁布墙面，搭配装饰匾、黑白挂画、白色的枯枝等装饰品。有的包间还设计了一个八角形的门框，使中式的韵味更为突出。在VIP包间还规划了一处景观的意境，使空间更具有情调。

　　卫生间的墙面和地面采用了水泥砖，做旧的木作台盆、做旧的木拼条隔断与水泥砖搭配十分协调。

Water landscape area adopts simple design and basic elements, such as plain cement walls, white projection, pottery, water features etc, forming a natural and poetic picture.

The corridor uses simple cement boards and wood batten, showing the pursuit of surface texture. In order to weaken toughness of the material, the designer also cares about details. For example free placement of dead wood and stone landscape at the end of the aisle, the pots and tea sets in the cabinets, black and white paintings on the wall, all of these designs make the space look richer.

Rooms do not have complex shapes, dominated by plain color. Beige straw lacquer, white wall cloth are the main decorative elements, supplemented with black and white paintings, white dead wood etc. Some of the rooms are installed an octagonal door frame, full of chinese flavor. The VIP room adopts an artistic conception of landscape, which makes the space more flavor.

Bathroom wall and floor use cement bricks, old timber basins and wood partitions.

天域阁
Horizon Club

项目名称：天域阁
项目地点：新疆乌鲁木齐
建筑面积：1050 ㎡
主要材料：黑色钢艺雕花、木地板、瓷砖、花砖
主案设计：蒋国兴
空间摄影：吴辉（牧马山庄空间摄影机构）

Project name: Horizon Club
Project location: Urumqi, Xinjiang
Building area: 1050 ㎡
Main materials: Black art carved steel, wood flooring, tiles, tiling
Project designer: Jiang Guoxing
Photograph: Wu Hui (Graze horse villa photograph organization)

天域阁坐落于乌鲁木齐骑马山，建筑外观为欧式风格。为了配合项目的整体定位，设计师将室内设计与建筑外观做了完美的融合，整体以欧式风格为主，局部混搭传统民族风格。餐厅的整体视觉感受，定位于"大气中不失细腻、不经意间流露出高贵品质"。天域阁带领叙品开启设计的新旅程，不同于以往的现代中式风格，以欧式风格与民族风格的结合为顾客呈现一场视觉盛宴。

Horizon Club is located in Qima Mountain, Urumqi. The architecture is European style. In order to fit the project's overall positioning, designer combines interior design and architecture appearance to make a perfect fusion of European style, which dominates the whole design, and mixes with traditional national style. Overall visual feel of the restaurant is positioned on the "graceful yet delicate, inadvertently revealing the noble qualities". Horizon Club has led Xupin start a new design journey and different from previous modern-chinese style, it presents by combination of European style and national style a visual feast for the guests.

　　优雅谦逊、低调内敛，这些习惯形容绅士的名词，都被拟人化地赋予一种色彩——高级灰。它深沉不张扬，却风度翩翩、彬彬有礼，犹如一位魅力绅士，信步向你走来。二楼的空间整体被赋予高级灰，服务台采用黑色钢艺雕花，底部透光，呼应银质吊灯，犹如西域蒙面的少女翩翩起舞。

　　二楼大厅为挑高空间，6米高的拱顶，被贴以金箔，投以灯光，瞬间将你拉回14世纪的文艺复兴时期。高大的穹顶庄重威严，让人肃然起敬。顶部一片云朵贯穿整个大厅，点缀酒红色、黄色、蓝色，让整个空间重回时尚行列。

　　二楼的散座区，顶部为蓝色与黑色结合的八边造型吊顶。静谧幽蓝的光，增添了神秘感，越发引人入胜。

Elegant and modest, these nouns used to describe a gentleman, was anthropomorphically given a color – advanced gray. The features of this color are deep quiet, personable, courteous, as a charming gentleman, walking towards you. The whole second floor use advanced gray. The service desk uses black art carved steel and the bottom light echoes silver chandelier, like a western masked girl dancing.

The hall on the second floor is an elevated space, with a 6-meter-high ceiling covered with gold foil. When casting lighting, it will instantly pull you back to the 14th century, during the Renaissance. The tall dome looks solemn, deserving respect from people. At the top of the hall, a cloud goes through the whole space. By decorating red, yellow, blue elements, the hall looks modern.

Circle area is also on the second floor, the top of which is a blue and black octagonal suspended ceiling. Quiet blue light adds a sense of mystery and is more engaging.

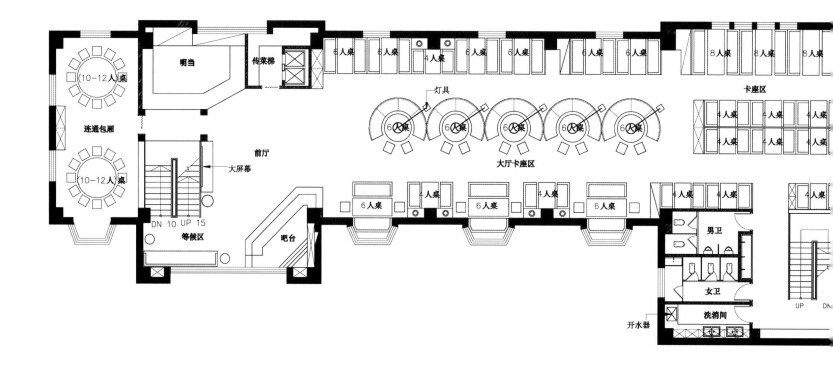
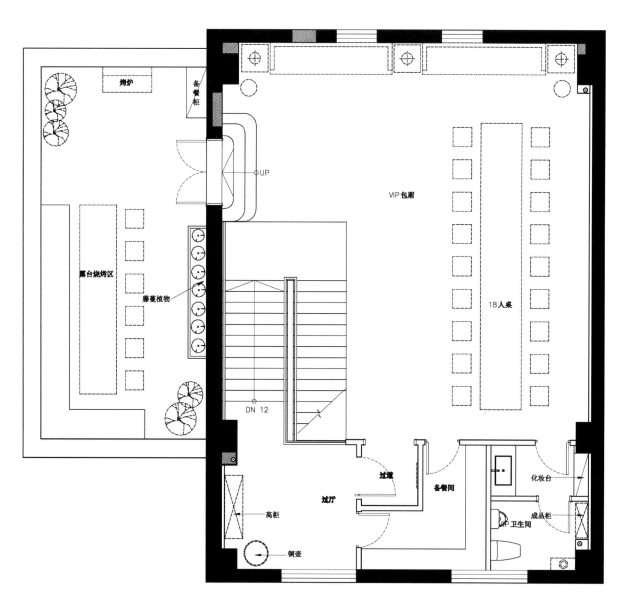

  二楼的左边是竹子装饰的景观，木拼条装饰着服务台及整个走道。走道没有多余的光源，三条回字形的灯片把过道分成一段一段的。在每个暗门的入口处都悬挂了一个锈铁板雕刻的门牌，一束束光源照射在每个门牌上。在走道的尽头，一朵云灯点缀了整个过道，墙面的镜子则拉伸了空间感。

  包间没有复杂的造型，以素色为主。米黄洞石，鹅卵石，灰色木拼条，灰色火山岩装饰的墙面，再搭配竹篱笆、黑白挂画、枯枝等装饰品。在每个包间都规划了一处竹子景观，使空间更具有情调。

  卫生间延续了走道的元素，木拼条装饰的墙面和黑色的地面使空间看起来更质朴。

Bamboo decorations are set on the left side on the second floor. Wooden battens decorate service counter and the whole corridor. There are not extra light on the corridor, three "回"-shape (a shape of Chinese word) lights separate the space of corridor. At the entrance of every hidden door, there is a doorplate made of rusted board, each light shines onto each doorplate. At the end of the corridor, a cloud lamp embellishes the whole corridor, mirrors on the wall expand the sense of space.

There are not complex shape in the private room. The decoration there focus on plain-style. Cream travertine, cobbles, gray wooden battens, and gray volcanic stones decorate walls. They are matched with bamboo fencing, black and white paintings, and deadwoods in harmony. There is designed a corner for bamboo decorations in every private room, that makes space atmosphere appealing.

The washroom extends elements of the corridor. The wall decorated by wooden battens and the black floor make the whole space natural and concise.

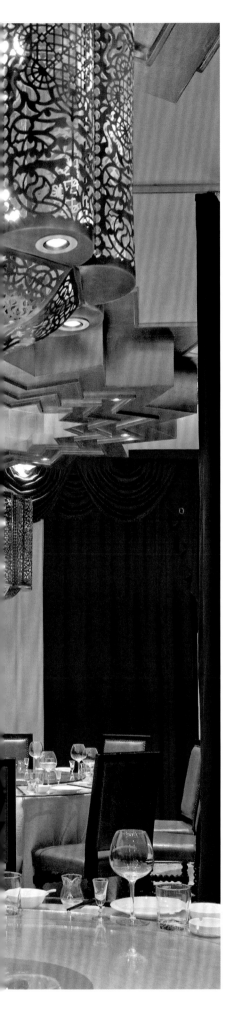

花枝沸腾鱼
Flowers-boiled Fish

项目名称：花枝沸腾鱼
项目地点：新疆乌鲁木齐
建筑面积：850 ㎡
主要材料：木饰面、木格、石材荔枝面处理、灰镜、木地板、海藻泥
主案设计：蒋国兴
空间摄影：吴辉（牧马山庄空间摄影机构）

Project name: Flowers-boiled Fish
Project location: Urumqi, Xinjiang
Building area: 850 ㎡
Main materials: Timber veneer, wood grid, bush-hammered stones, grey mirror, wood flooring seaweed
Project designer: Jiang Guoxing
Photograph: Wu Hui (Graze horse villa photograph organization)

本案位于古丝绸之路新北道上的重镇乌鲁木齐，简约的现代时尚感与东方元素的抽象剥离深植于整个空间之中，蕴含大气深邃的东方意境。设计师从传统东方元素中汲取灵感，如精致典雅的青花瓷盆，韵味十足；入口处有质感的海藻泥与中式木隔断虚实对比张弛有度。大厅内一座座鸟笼形木隔断相对独立私密而又独具个性。包间走道一排粗犷大气的中式金色门框，颇有皇家气派。设计师炉火纯青地运用厚重色彩，配以白描挂画使得空间大气、生机勃勃。这些传统的东方文化，元素绝不是简单地罗列，而是通过当代设计形式、语言，张扬地表达当下的审美气质。在这个充满想象的空间里，当代艺术和传统文化邂逅，艺术与空间碰撞，使空间充满活力，并拥有一种浪漫主义的气息。

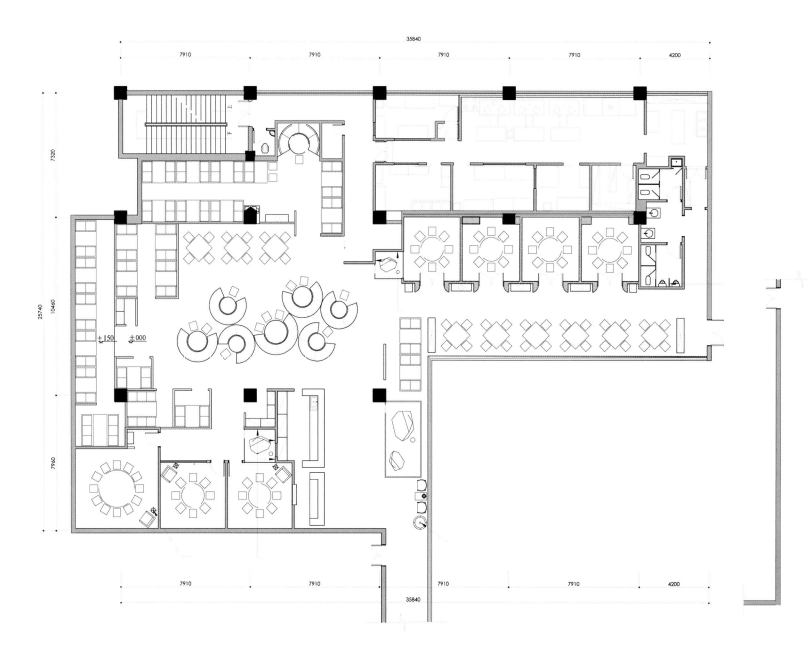

This project is located in the ancient Silk Road, New North Road, Urumqi. Simple modern style and oriental elements are deeply rooted throughout the space, with profound oriental flavor. Designers draw inspiration from the traditional oriental elements, such as refined and elegant blue and white pottery basin. Seaweed and Chinese style wooden partition at the entrance make a clear comparison. A cage-shaped wooden partition is placed at the corridor, independent and unique. A row of and Chinese-style golden door frame make the space look great. Designers perfected the use of heavy color, and watercolor paintings are hanged in my rooms. These Oriental cultural elements are well used to show the design for modern times. In the space full of imagination, modern art meets traditional culture, which makes the whole room full of vitality, restful and romantic.

唐人街雅足轩
Chinatown Elegant Foot Porch

项目名称：唐人街雅足轩	Project name: Chinatown Elegant Foot Porch
项目地点：新疆乌鲁木齐	Project location: Urumqi, Xinjiang
建筑面积：2600 ㎡	Building area: 2600 ㎡
主要材料：仿木地板砖、藤编壁纸、稻草漆、火山岩、水泥砖	Main materials: Wood floor brick, rattan – weaven wallpaper, straw lacquer, volcanic rock, cement brick
主案设计：蒋国兴	Project designer: Jiang Guoxing
空间摄影：吴辉（牧马山庄空间摄影机构）	Photograph: Wu Hui (Graze horse villa photograph organization)

中国足疗历史源远流长，西汉《礼记》中记载了以中草药熏、蒸、浸、泡的疗法。"足是人之根，足疗治全身"，古时候扁鹊根据人的生活习惯，发明了用中草药煎汤泡脚的祛病方法，据说这就是中药足浴、足疗的前身。我国是足部疗法起源最早的国家。中国几千年前就有关于足部治疗的记载，据考证，在医学经典著作《黄帝内经》里就有关于足部治疗"观趾法"的记载。《史记》中也详细地记载了"足心道"。

大唐盛世，唐朝为中国五千年历史中鼎盛的朝代之一，宋代的艺术文化也是达到鼎盛时期。而足部疗法在唐宋时期传入日本、朝鲜后走向世界。本案风格为唐风宋韵，取自中国最具代表的两个朝代，借此表达本案的宽度、广度和包容度。

In China, pedicure culture has a long history, it has recorded since the Han Dynasty. The Book of Rites recorded in detail some therapies with herbs like frying, smoking, steaming and soaking. "foot is the root to body, pedicure is of benefit to health". In ancient China, the doctor Bian Que found the cure by soaking feet with hot water which boiled herbs according to people's living habit. It is said that the cure is the embryo of pedicure. China is the earliest country where people begin to keep fit by pedicure. it has recorded about foot massage since thousands of years ago in China. It is a fact that there are many foot therapies in the classic Chinese medical book- "Huangdi Neijing" and the therapies like "observe foot", "the cure of sole" are recorded in Records of the Historian.

Tang Dynasty is regarded as an affluent era in history, and it is one of the most important and prosperous dynasties in three thousand years in China. In Song Dynasty, the attainments of art culture had arrived the top. It was at this time that pedicure spread into Japan, Korea and the world. This project adopts Tang & Song Dynasties style, which is derived from the two representative dynasties to express the inclusion.

　　本案立足于人的需求，倾力创造一个轻松舒适的中式休闲空间。没有过多的色彩修饰，没有大量的造型堆叠，一切还原材质本身的面貌，清新脱俗，不染红尘。光线、气味、声音是稍纵即逝的，而设计师却通过对触觉、视觉、嗅觉的切割、糅合，带你进入一个极具魅力的新国度。

　　在空间布置上，将健身区、办公区、展示区、接待区、娱乐区及茶室布置在一层，二层则以按摩足浴间为主。

　　大厅服务台，位于整个空间的中轴线上。外观造型上用木条编织的手法呈现，上下对称，左右居中。"中"是天地万物最神秘的一个点，静与动，爱与恨，进与退，大千世界的平衡往往就取决于这个微小的点，将服务台布置在这里，意图将最好的服务发散开来，平衡这里的一切事物。

The project is based on people's necessary, and is aimed to create a relax and comfortable Chinese-style leisure space. Without redundant color decoration and complex shapes, all materials embody its real look, the design makes a natural and pure feeling. Light, smell, and voice is temporary, but designer lead you in a new charming world by the mixture and separation of the tactile sense, vision, and smell.

In layout, fitness area, office, display area, reception area and tea room are designed on the first floor, and the space on the second floor are mainly designed for massage and pedicure.

The service counter in lobby is located in the centred line of the whole space. Sculptural decorations include weaving with battens, all elements are set in symmetric. Core is the most mysterious point of all things, the balance between love and hate, dynamic and static, advance and retreat all over the world are decided by the point. The position of service counter means the best service originates from there, and the service could make all things there be in a balance.

转角而遇的中式圆形内发光门洞，光影交织，错落有致，极致简约却又气度不凡。天圆地方，圆为天、为乾、为阳、为东。八卦之中乾卦为阳之首，乃万物之初始。这里运用圆来寓意，一切在这里开始，序幕才刚刚拉开。

圆形门洞层层铺开，光与影的交织，形与质的碰撞，景中有景，画中有画，可谓一步一景，处处是景。

展示区墙面木格栅与顶面的竹子编织相呼应，整个竹子编织使空间通透不沉闷，疏密有致，节奏得当。锈铁板则有种与生俱来的厚实感，历经时间的洗刷，虽已锈迹斑斑，但这又何尝不是经历风雨后另一种坚韧的美。

When you turn around and you can see the round and Chinese-style door opening which is lighted inside, the light and the shadow form a rhythmical image that is very concise and advanced. "the sky is round, the ground is square" says a Chinese proverb. The round represents the sky, the Qian, the Yang, the east. In Bagua (an ancient Chinese philosophy), Qian is Yang which is the beginning of all things. The round door opening here means everything is originated from here and the prelude just begins.

The round door opening spread in layers, the light and the shadow mix together, the feeling of shape pump into the feeling of texture. There are views in views, there are paintings in paintings. Everywhere you can find is the scenery.

In display area, wooden grids on the wall is matched with bamboo weave on the ceiling. The rhythmic Bamboo decorations relax the space. Rusted board is massive naturally, through a long time, it shows another kind of beauty.

书吧的墙面采用灰色瓦片堆砌，厚重古朴。

按摩室的顶部由竹条编织而成，别出心裁。光线透过间隙落下来，层层叠叠，让人着迷。简洁的线条，铿锵有力，使得空间内有了力量。精致的灯笼吊灯，点亮了房间，映着江景，屋内景色分外别致。

健身房的复古木地板，与墙面的木饰面，和谐统一。办公空间墙面使用黑色线条分割，配以藤编壁纸，干练之余，不至于拘谨。

茶室空间整体使用舒适的灰色调，如谦谦君子，低调内敛，毫不张扬。

棋牌室色彩对比强烈，灰绿壁布，配以红色装饰柜，别有韵味。卫生间条砖的线条和镜面的搭配增强了空间感，色调柔和，不突兀。

The wall in book bar is made of piled tiles, it is massive and rustic.

Bamboo weavings on the ceiling of massage chamber is special. Layer upon layer of Light penetrate through the gaps, the whole atmosphere is fascinating. Concise strips make the space powerful. Delicate ceiling lamps brighten the room, it looks more beautiful matched with river views.

The retro wood flooring in the fitness room is matched well with wood finish on the wall. Office wall is separated by black lines matched with rattan weaver wallpapers, the whole decoration looks concise but restrict.

The color tune of tea room is gray, which is like a modest gentleman.

The color in the card room is contrasting. Celadon wall clothes matched with red decorate cabinet has a special feeling. The lines of narrow bricks in the washroom matched with mirror expand the space in visual and the color tune is gentle.

一层卫生间，白色石材与木质隔板完美搭配，自然赋予的肌理和不同的质感，碰撞出不一样的精彩。

二层门廊内一条条光影交叠，直至远方。一旁的水景打破了沉闷，上善若水，水善利万物而不争。水是生命的起源，又为足疗的依托。水景为设计的点睛之笔，映衬在交织的瓦片旁，厚重、质朴感无不将你围绕。

通道两侧细密的木质格栅纤细稳健，抬头穿过细窄的天井，仿佛时间交错，恍惚间回到清代的楼阁，廊檐蔓回，如悠扬的歌声，久久于心，不能忘怀。

包间内，以光影为主要的表达形式，巧妙地将无形转为有形。整体为木色调，点缀红蓝色彩，将现代的时尚带入古朴的空间。

二层洗手间内的镜面如两滴流动的水珠，造型别致，富有动感。墙面材质为火山岩，凹凸不平的肌理质感，有一种原生态的美，卫生间与洗手间被水泥砖墙面统一起来，错落有序的拼贴，将整个空间合二为一。

White stones in the washroom on the first floor is matched well with wooden board. Natural textures has different feeling and different beauty.

On the second floor, layer upon layer of light extend to the distant place on the hallway. Water decoration nearby break the silence of the space. Greatest kindness is like water, water brings all benefits to all kind of things without getting payback. Water is the source of life and it is also the basic thing of pedicure, so the water decoration is a smart and appropriate setting which makes the piled tiles nearby more massive and rustic for the whole atmosphere.

On the both sides of corridor, slim and dense wooden grills is stable. When you look up through narrow skylight, it seems to go back in Qing Dynasty, you were on the corridor of attic, and you couldn't forget sweet music echoes around for a long time.

In private room, decoration of light and shadow is the main expression which become abstraction to concreteness. The color of red and blue that embellished in the whole color tune of wood change your feeling from ancient to fashionable.

On the second floor, two mirrors of washroom are like two drops of flowing water whose shape is unique and dynamic. The decoration materials on the wall are wild and beautiful, volcanic stones whose surface are rough. Two spaces of washroom and toilet are uniformed by the same materials of cement bricks which is arranged regularly.

君元恒基
June Henderson Office

项目名称：君元恒基
项目地点：新疆乌鲁木齐
建筑面积：660 ㎡
主要材料：爵士白大理石、做旧木色木地板、做旧本色木饰面、
驼色藤制壁纸
主案设计：蒋国兴
空间摄影：蒋国兴

Project name: June Henderson Office
Project location: Urumqi, Xinjiang
Building area: 660 ㎡
Main materials: Jazz-white marble, antique finish
wood flooring, wood veneer,
camel color rattan wallpaper
Project designer: Jiang Guoxing
Photograph: Jiang Guoxing

办公室通常是从事脑力劳动的场所，办公的环境影响着每一位员工的工作情绪和工作效率。本案内，白色的大理石、木本色的装饰、黑白色搭配的家具、驼色的藤编壁纸、通透明亮的木框玻璃隔断和走道处别出心裁的景观，无不传达出一种轻松、愉快的感觉，更体现了 种和谐的企业文化。水吧区、等待区、洽谈区的设置使空间规划更合理，而且更具人性化，充分考虑员工及客户的需求，营造了一种健康、舒适、以人为本的办公环境。

在灯光的运用上，没有过多的复杂光源。简单的格子顶面配以明亮的灯光，再加上自然光源的照射，使整个空间干净而明亮。

Office is usually taken as the place for mental work, the environment of which affects the mood and the efficiency of every employee. White marble, wood-colored decorations, black and white furniture, camel color rattan wallpaper, transparent and bright wood-framed glass partitions, unique landscape, all convey a feeling of relaxed and happy feeling, embodying the harmonious corporate culture. Water bar, waiting areas and discussion area make the spatial planning more reasonable and more user-friendly, fully consider the needs of employees and customers, and create a healthy, comfortable, people-oriented work environment.

In the use of light, there are not too many complex light sources. Simple grid ceiling with bright light and with exposure to natural light, makes the whole space clean and bright.

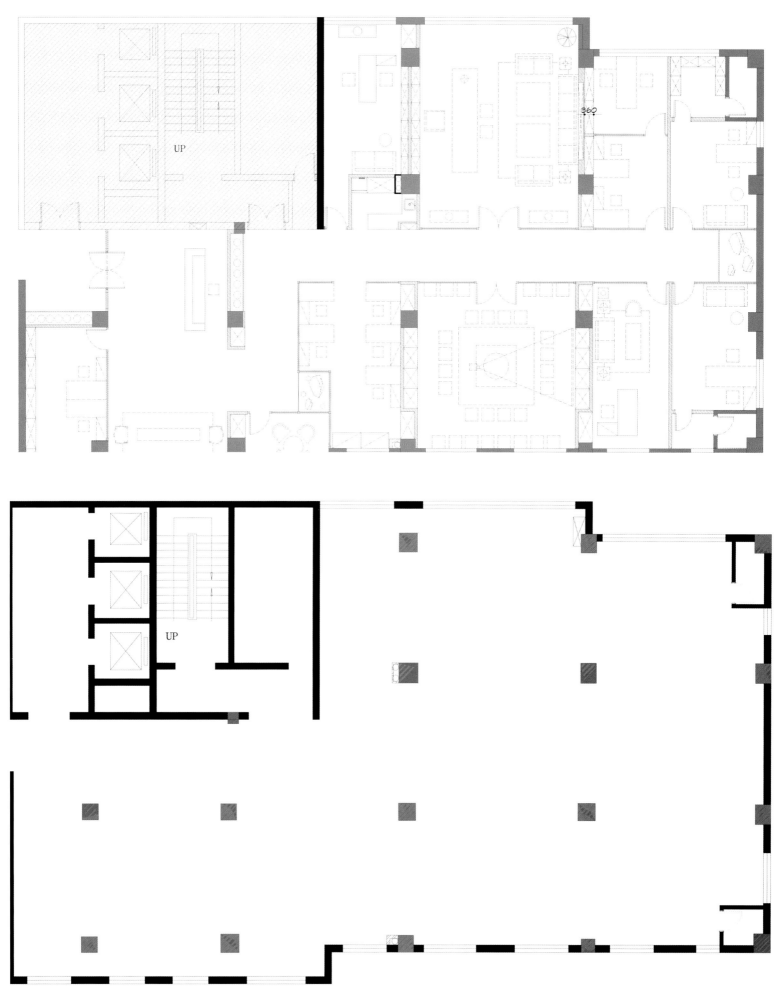

本案是一个面积大约为600平方米的办公空间。在功能布局上，充分考虑了通风、采光、私密的需求，把每个办公室规划在靠窗的位置，把董事长、副总、财务办公室规划在公司整体的后方。

This case is about 600 square metres. On the functional layout, designers take fully into account the needs of ventilation, lighting, privacy, placing the seats near the windows, and the offices of the company's leaders are set at the back of the inner place.

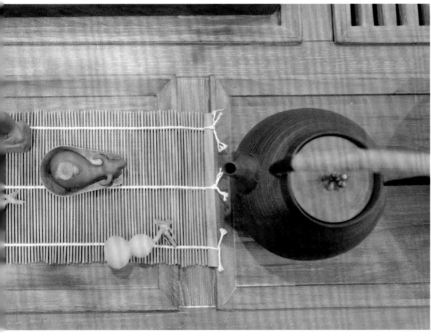

原膳
Raw Food

项目名称：原膳
项目地点：新疆乌鲁木齐
建筑面积：5000 ㎡
主要材料：黑色花岗岩、木饰面、文化石、浅灰色壁纸、深色木地板
主案设计：蒋国兴
空间摄影：蒋国兴

Project name: Raw Food
Project location: Urumqi, Xinjiang
Building area: 5000 m²
Main materials: Black granite, wood finishes, culture stone, light grey wallpaper, dark wood floor
Project designer: Jiang Guoxing
Photograph: Jiang Guoxing

"物华天宝丝绸路，人杰地灵丁古城"。本案位于我国西部广阔而又富饶的宝地新疆首府乌鲁木齐。古丝绸之路把承载着东方传统文化精髓的陶瓷、丝绸、漆器带到这片圣土。而今浮华与喧嚣的环境中东方禅意在这里找到一片灵魂的栖息之地。

"The historical Silk Road, an outstanding historic city". The project is located in the broad and fertile land in Urumqi, the capital of Xinjiang. Ancient Silk Road brought pottery, porcelain and lacquer ware to this place. Now, the designer wants to create a space, where people can relax in Zen atmosphere.

　　本案设计师采用深色基调营造东方禅意。用简洁硬朗的直线条勾勒极具层次感的空间；以朴实的手法，通过虚与实、明与暗、简与繁的辩证结合实现一种"古韵新风"。设计师并不拘泥于细节的刻画，而是用大块的面带过，并恰到好处地用一些中式元素来烘托意境，给人更多的留白空间去品味思索。东方主题的艺术陈设可以淡雅如君子，可以贵气如帝王，可以轻描淡写，又可以浓妆重抹，各具风骚。入口处富有创意的白色同心圆背景墙从视觉上给人强烈的冲击感，错落有致的流水鸟笼造景、大气古朴的传统宣纸灯具、清雅黯然的陶瓷漆器制品、宁静致远的白描挂画，这些元素勾勒的景象，乍看若有若无，却让人难以忘怀；超大面积的包厢，配以简洁大方的现代明式家具，如意味深远的山水枯木造景、透出淡定的奢华黑色皮质沙发。整个包厢简约宁静，让人陶醉，似在诉说男人的情怀。以禅的风韵来诠释室内设计，不求华丽，旨在体现人与自然的沟通，以求为现代人营造一片灵魂的栖息之地。在这里或就餐饮酒，或交友会客，或商务洽谈都从容惬意，还可以体验一种新的生活方式。

Designers use dark tone to show Zen spirit in the case. Simple and sturdy straight lines outline the rich layering of space, through the comparison of virtual and real, light and dark, simple and complex dialectic realizing a "modern space with classic design elements". The designers don't get bogged down in details, but with large blocks, they use some Chinese elements to set the mood, giving more space to think. Oriental-themed art furnishings can be simple or elegant. The white background wall at the entrance is striking. The use of traditional elements, such as artificial landscape, lanterns made of rich paper, pottery and lacquerware, paintings, outlines an unforgettable scenery. Private rooms of big size use simple and generous modern Ming Dynasty style furniture, creating a quiet and relaxing space. The project uses Zen charm to explain Interior design, aiming at embodying the communication between man and nature, and creating a place for people to enjoy, where they can take meals, meet friends or have business meetings.

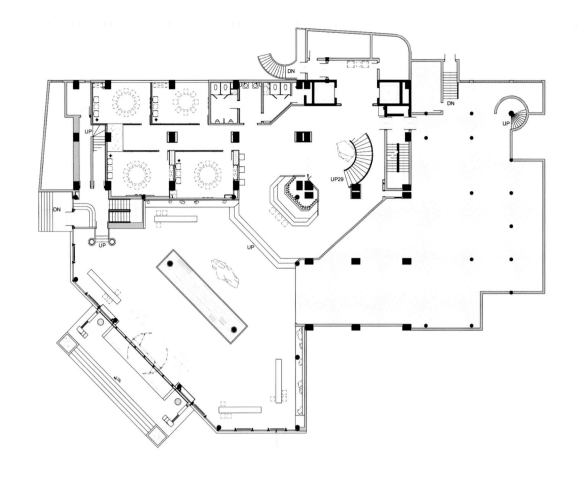
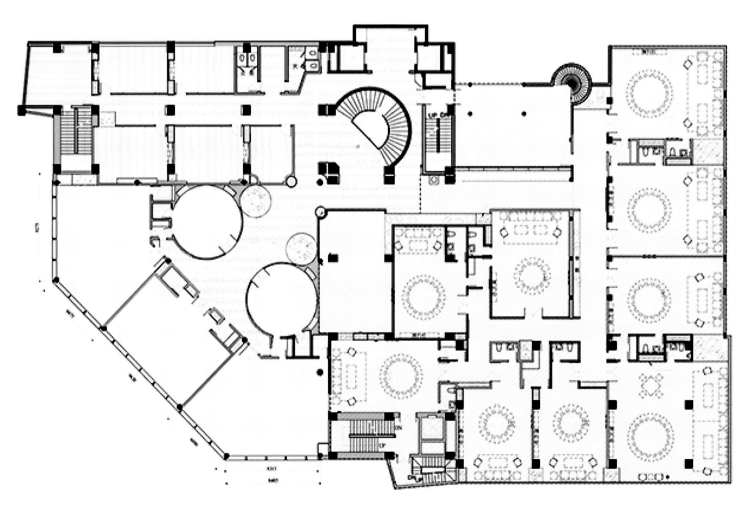

云峰投资公司
Yunfeng Investment Company

项目名称：云峰投资公司
项目地点：新疆乌鲁木齐
建筑面积：230 ㎡
主要材料：红砖、红瓦、水泥板、爵士白大理石、原木色实木板、藤编壁纸
主案设计：蒋国兴
空间摄影：吴辉（牧马山庄空间摄影机构）

Project name: Yunfeng Investment Company
Project location: Urumqi , Xinjiang
Building area: 230 ㎡
Main materials: Red bricks, red tiles, cement slabs, jazz-white marble, natural color solid wood panels, ratten-weaven wallpaper
Project designer: Jiang Guoxing
Photograph: Wu Hui (Graze horse villa photograph organization)

　　本案设计的是投资公司的办公部，空间不大，200多平方米。设计师意在给金融公司一个不一样的感觉，在色彩的搭配上选择了质朴的黑色加做旧的木色，点缀一些古朴的红色。

　　进门入口就是一面红色的九孔砖墙，质朴的红色，一块不规则的实木做的牌匾上面，铜质的"云峰投资公司"几个字凸出在木板上，旁边放一个大大的陶罐，插满了鲜黄色的花。天花上有两盏光圈射灯，光线打在牌匾和花上，使得这两处显得更加醒目。

This case is on the design of an investment company's office, over 200 ㎡. Designers aim at creating a different feeling for the company, and in terms of color, the designers choose to use rustic black, distressed wood and quaint red.

At the entrance stands a red wall of nine-hole bricks, the name "Yunfeng Investment Company" is mounted on a solid wood board, under which sets a pot full of flowers. Two spotlights are hanged on the ceiling, shining on the board and flowers and in turn attracting people's attention.

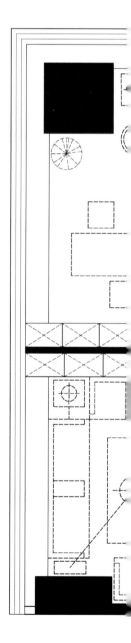

入门的四周墙面都是红色的小砖与瓦片错叠拼接而成，中间用黑色拉丝不锈钢线条分割，精致又有点古老的味道。地面是黑色的光面砖，像是一片黑色的海洋，倒映着暗红色的墙面，空间显得干净温暖。

Surrounding walls is made by splicing of red bricks and tiles, in the middle of which are divided by brushed stainless steel with black lines, exquisite and antique. Black smooth brick on the ground, like a black sea, reflects the dark red walls, making the space clean and warm.

走进去是个接待水吧台，用爵士白大理石做成长长的台面，边上是黄色的灯带，再放几把实木的椅子。吧台上面是镜子做成的层板，摆放一排仿真绿植，给空间增添一些绿色清新的气息。吧台的右边是一排矮柜。两盏黑色的小吊灯增添了空间的温馨气氛。

吧台后面用镜子和黑色不锈钢条做成宽窄不一的格子，跟红色的砖墙形式一样。镜子可拓展不宽敞的空间，使空间显得更有深度。走道中间有一个小小的景观，用白桦树和造型简单的石头做一道景墙，小巧精致。走道的尽头是用木拼条做成的，很巧妙地把门隐藏起来。

The reception bar counter is long with a work top made of white marble, yellow light belt and wood chairs next to the reception bar counter. Artificial plants are used to add fresh air to the space. On the right of the counter, there is a row of low cabinets under two ceiling lamps.

The back of the bar is a lattice frame made by splicing of mirrors and black stainless steel bars, the same style with the red wall. Mirrors can be used to broaden the space. In the middle of the corridor, a small landscape is created through a birch and stones, which looks beautiful and delicate. At the end of the corridor, a wall made of wood battens weakens the existence of the door.

多功能室的整个墙面用木拼条做成,有种做旧的感觉。沙发休闲区做了一面背景墙,用木色层板做成柜子,塞满了大大小小的圆木做装饰,错落地摆着陶陶罐罐和书本,再配上一台古老的留声机,装饰的意味就浓了起来。

其他的办公室用米色的藤编壁纸,原木色的压条、爵士白的大理石地面,让空间的色调偏于明亮。

The wall of the multi-function room is made of wood batten with an antique feel. At the sofa lounge area, the background wall and a wood cabinet are adopted. The cabinet is set with logs, jars, books and an old gramophone, forming a strong decorative meaning.

Other offices take a use of cream-colored rattan weaven wallpaper, wood color board, and white marble flooring, making the space bright.

松山行足道
Songshan Xing Foot Massage

项目名称：松山行足道
项目地点：新疆乌鲁木齐
建筑面积：3185 m²
主要材料：木地板、藤编壁纸、斧刀石、木格栅、水泥砖
主案设计：蒋国兴
空间摄影：吴辉（牧马山庄空间摄影机构）

Project name : Songshan Xing Foot Massage
Project location: Urumqi , Xinjiang
Building area: 3185 m²
Main materials: Wooden floor, ratten weaven wallpaper, cultural stones, wooden grid, cement bricks
Project designer: Jiang Guoxing
Photograph: Wu Hui (Graze horse villa photograph organization)

在中国的历史长河中不乏名人靠足浴养生保健的故事：唐朝杨贵妃经常靠足浴来养颜美容；宋朝大文豪苏东坡通过足浴来强身健体；清代名臣曾国藩更是视"读书""早起"和"足浴保健"为其人生的三大得意之举；近代京城名医施今墨也是常用花椒水来泡脚养生。可见足浴在中华养生保健历史中占有很重要的地位。

There are many stories of famous people about health care in the long course of Chinese history. The imperial concubine Yang in Tang Dynasty often chose pedicure for beauty. The great litterateur Su Dongpo in Song Dynasty chose pedicure for health. The famous statesman Zeng Guofan in Qing Dynasty thought that the pedicure was as important as reading and getting up early in his life. The famous doctor Shi Jinmo in Beijing in modern time also chose footbath with pepper-water for health constantly. The above examples suffice to show that pedicure plays an important role in the Chinese health-care history.

大厅墙面运用竹编与黑色压条有序地分割。吧台背景采用斧刀石，自然的肌理，天然的质感，一切回归自然。吧台运用鱼鳞状格栅，造型在灯光的照射下，显得格外耀眼。还运用了红色的瓦砖点缀，让色彩变得鲜亮明快。

整个公共区域地面采用黑金沙、山西黑、中国黑三种颜色石材拼花，体现一种沉稳、低调。门采用的是木拼条通顶的设计手法，延伸了空间尺度。过道墙面采用水泥砖用原木色压条分割，其中还有竹林景观，灯光透过竹林，使得整体空间活泼而有序。

There are two attractive sights in the hall, one is the wall paved by black bamboo weaving orderly, the other one is the counter of the bar, made of rough stones with fish scale shaped grid, shinning in the top light.

In the public area, the floor is paved by the marble of Black Galaxy, Shanxi black and China black, steady and peaceful. The door is made of full-height wood battens to extend the space, and the wall is paved by cement bricks with wood battens to divide the space, light comes through bamboo forest, making the space lively and orderly.

包间顶部采用竹条编织，别出心裁。其简洁的线条，铿锵有力，使得空间内有了力量。精致的灯笼吊灯，点亮了房间，映着江景，分外别致。墙面用水墨风格的装饰画点缀，写意不写实，给人以超脱万物、置身于仙山灵水的感觉。最简单的魅力所在，更高深的东西，只能靠自己去摸索、去体会……

茶室设计取自中国古文化，横平竖直，方方正正，并设置竹林景观，因为竹枝杆挺拔修长，亭亭玉立、四时青翠、凌霜傲雨，竹子具有"宁折不弯"的豪气和"中通外直"的度量，它性质朴而淳厚，品清奇而典雅，形文静而怡然，正所谓"未出土时已有节，待到凌云更虚心"。身处在这一片静谧的竹林中，不忍大声喧哗，怕惊扰这美好。

The light leaking from the ceiling of bamboo weaving looks charming and attractive, and it also powers the space with some simple lines. Against the river view, lantern-shaped ceiling lamps light up the room in the night. Walls are decorated with Chinese traditional painting, through which the guests will find their own ways to experience a feeling around mountains and waters.

The idea of tea room design origins from Chinese ancient culture, taking a square shape. In China, bamboo is rather a style of life than a plant. It's thin but long and straight, and green all the year wound. The character of "rather break than bend" always describes the character of a person. And so, the landscape of bamboo forest bring here peace. In this place, everyone would not like to speak aloud and is afraid of breaking the peace.

红酒吧顶面采用镜面马赛克，配合灯光的运用，低调奢华的氛围，让人感觉更多的是神秘色彩。

洗手间延续了过道的设计手法，体现出整体性、统一性。卫生间墙面使用黑色线条分割，配以竹节砖，竹节有着节节高升、步步高升的寓意。

竹，秀逸有神韵，纤细柔美，长青不败，高风亮节，高尚不俗，生机盎然，蓬勃向上……它有许许多多的优点，然而，我最欣赏它的坚贞不屈，它那"孤生崖谷间，有此凌云气"的美好品质。它柔中有刚的高尚品德时时刻刻激励着我！

In the bar, the ceiling of mosaic mirror reflects the light to many directions, which make the atmosphere luxury and mystery.

In the washroom, the decoration extends the design of the aisle to keep the overall unity of the whole space. The wall is split by black line and paved by bamboo tiles meaning getting better in life and career.

Bamboo is especially appreciated by people because of its slender, evergreen, ethical, noble, vibrant, vigorous and a lot of advantages. Of all, I appreciate its character of unyielding. A poem is to describe the trait of bamboo: "It grows in valleys, but its ambitions soar to the sky". That's why bamboo elements are often seen in my designs.

图书在版编目（CIP）数据

　　叙品十年：蒋国兴作品集 / 蒋国兴主编. -- 南京：
江苏凤凰科学技术出版社，2017.1
　　ISBN 978-7-5537-7429-9

　　Ⅰ．①叙… Ⅱ．①蒋… Ⅲ．①饮食业－服务建筑－室
内装饰设计－作品集－中国－现代 Ⅳ．①TU247.3

　　中国版本图书馆CIP数据核字（2016）第272976号

叙品十年　蒋国兴作品集

主　　　　编	蒋国兴
策　　　　划	徐宾宾
责 任 编 辑	刘屹立
特 约 编 辑	艾　璐
出 版 发 行	凤凰出版传媒股份有限公司
	江苏凤凰科学技术出版社
出版社地址	南京市湖南路1号A楼，邮编：210009
出版社网址	http://www.pspress.cn
总 经 销	天津凤凰空间文化传媒有限公司
总经销网址	http://www.ifengspace.cn
经　　　　销	全国新华书店
印　　　　刷	上海利丰雅高印刷有限公司
开　　　　本	965 mm x 1 270 mm　1/16
印　　　　张	19.5
字　　　　数	156 000
版　　　　次	2017年1月第1版
印　　　　次	2017年1月第1次印刷
标 准 书 号	ISBN 978-7-5537-7429-9
定　　　　价	298.00（精）

图书如有印装质量问题，可随时向销售部调换（电话：022-87893668）。